In

Se
Mi ⌐alifornia, Santa Barbara, and Paul McDonald,
Un ...n, UK

Th. ...national Screen Industries series offers original and probing analysis of media industries around the world, examining their working practices and the social contexts in which they operate. Each volume provides a concise guide to the key players and trends that are shaping today's film, television and digital media.

Published titles:
The American Television Industry *Michael Curtin and Jane Shattuc*
Arab Television Industries *Marwan M. Kraidy and Joe F. Khalil*
East Asian Screen Industries *Darrell Davis and Emilie Yueh-yu Yeh*
European Film Industries *Anne Jäckel*
European Television Industries *Petros Iosifidis, Jeanette Steemers and Mark Wheeler*
Global Television Marketplace *Timothy Havens*
Hollywood in the New Millennium *Tino Balio*
Latin American Television Industries *John Sinclair and Joseph D. Straubhaar*
Video and DVD Industries *Paul McDonald*

Forthcoming:
The Indian Film Industry *Nitin Govil and Ranjani Mazumdar*
Latin American Film Industries *Tamara Falicov*

The Video Game Business

Randy Nichols

A BFI book published by Palgrave Macmillan

For my parents, who taught me to ask questions and never once wished for me to stop

First published in 2014 by
PALGRAVE MACMILLAN

on behalf of the

BRITISH FILM INSTITUTE
21 Stephen Street, London W1T 1LN
www.bfi.org.uk

There's more to discover about film and television through the BFI.
Our world-renowned archive, cinemas, festivals, films, publications and learning resources are here to inspire you.

Palgrave Macmillan in the UK is an imprint of Macmillan Publishers Limited, registered in England, company number 785998, of Houndmills, Basingstoke, Hampshire RG21 6XS. Palgrave Macmillan in the US is a division of St Martin's Press LLC, 175 Fifth Avenue, New York, NY 10010. Palgrave Macmillan is the global academic imprint of the above companies and has companies and representatives throughout the world. Palgrave® and Macmillan® are registered trademarks in the United States, the United Kingdom, Europe and other countries.

Cover images: (front) © Shadowadrik; (back) © Randy Nichols

Set by Cambrian Typesetters, Camberley, Surrey
Printed in China

This book is printed on paper suitable for recycling and made from fully managed and sustained forest sources. Logging, pulping and manufacturing processes are expected to conform to the environmental regulations of the country of origin.

British Library Cataloguing-in-Publication Data
ISBN 978–1–84457–317–2 (pb)
ISBN 978–1–84457–318–9 (hb)

Contents

Acknowledgments

To say that this book has been a long time coming is an understatement. Along the way, this book has seen a change of publisher and my own changes of universities. As such, there are far too many people to name who deserve thanks for helping shepherd it along the way. My thanks to the various editors at BFI Publishing and Palgrave who helped the process. It owes much to the research assistance from librarians at three universities – Cleveland State University, Niagara University and Bentley University – for helping me unearth some of the harder-to-find global data. As the process has drawn on, my friends and colleagues in the Southwest Popular/American Culture Association, the Union for Democratic Communications and the International Association for Media and Communication Research who have heard versions of chapters, helped refine conclusions, gathered international data and offered support. Particular thanks to Judd Ruggill, Ken McAllister, Ben Aslinger, Nina Huntemann, Matt Payne and Aphra Kerr on the game studies side of things for their feedback and suggestions along the way (even when they may not have known they were talking about this project), and to Janet Wasko, David Hesmondhalgh, Peichi Chung, Micky Lee and Eileen Meehan on the political economic side for their help in thinking through how to best contextualise the global industry. Finally, thanks to my great friends from graduate school and elsewhere Tricia Brick, Nicole Schuller, Joe Bell and Shaun Kohn for talking me off more than a few ledges and for their continued support.

Introduction: Video Games as a Global Phenomenon

Ten years ago, a book attempting to justify video games as a unique cultural industry would have been a tough sell. The common view was that they were just toys – admittedly fancy, costly toys, but toys nonetheless. Today, however, that view has changed. Video games have emerged as a fully developed industry that not only competes with but also often complements a wide variety of other cultural industries. Educators and academics have recognised their potential value as teaching tools since the first prototypes were developed in the 1960s. Similarly, concerns over potential negative effects, whether from too much time spent with video games or from exposure to violent content, have dogged the industry since at least the early 1980s.

In 2004, the global video game industry first earned more than Hollywood's domestic box office, a marker often used to demonstrate its growing importance. Moreover, a number of major media and communication companies began to use video games as integral parts of branding and advertising, including franchises such as *Harry Potter*, *Lord of the Rings*, *James Bond* and the NFL (Bloom, 2001; Bloom and Graser, 2002). Even popstar Britney Spears has a video game. The industry, which has tried to follow the Hollywood film system throughout its history, has even created its own hall of fame and the 'Walk of Game' in San Francisco (Harris, 2005).

One doesn't have to look far for examples of how significant and deeply ingrained video games have become in our day-to-day lives. They have been involved in a political scandal, as when a Norwegian representative was caught playing one during a major policy debate (CNN, 2003). Special 'serious' video games have been created for a variety of purposes. Serious games, which use the mechanics and capabilities of video games for training, have been used in a variety of areas including a number of political campaigns in the US. In fact, the ability of video games to energise the youth vote has led some experts to predict they'll soon become a mainstay of most political campaigns (Foster, 2004b).Video games are being used as a part of medical treatment and have been the focus of debates about youth violence (AP, 2005b, 2005c; ESA, 2001; Johnson, 2004; New Zealand Herald, 2003). Industry studies show that video games are less and less toys for kids (ESA, 2002, 2004, 2005a, 2006, 2007). Increasingly, they're facing the same questions of intellectual property ownership and piracy as the recorded music and film industries (Chazan, 2005; Veiga, 2004).

Moreover, the pedagogical value of video games has become so widely accepted that a school district in Michigan loaned PlayStation 2s to students to help them take advantage of educational software (Laskowski, 2005). Even the US military has shifted its long-standing use of video games into overdrive. The US Marine Corps has begun to use video game first-person shooter *Doom* (1993) to help teach its recruits tactics (McCune, 1998). Similarly, the US Army spent millions to develop and market its own video game, *America's Army* (2005), to inform potential recruits about army life (Brickner, 2004; Nichols, 2009; Wadhams, 2005). Other groups have taken note, with Hezbollah creating two video games to help bolster its cause, while Damascus-based Atkar Games has created two games geared towards countering portrayals of Muslims in most Western-designed video games (Nichols, 2009). Most tellingly, an increasing number of universities around the world offer courses in video game studies and even degrees in video game design (Barlett, 2005; Carlson, 2003a; Foster, 2004a).

Work in the video game industry has come to represent a major new hope for professional training in universities. Jobs in the industry have been described as 'some of the best jobs the American workplace has to offer' (Richtel, 2005b). In spite of this, increasing numbers of employees are leaving the industry or filing lawsuits citing unfair labour practices (ea_spouse, 2004a, 2004b). Understanding how the industry works is of paramount concern to video game studies, as education in the field is becoming professionalised.

But there are other reasons a detailed examination of the industry is needed. Understanding the forces that produce video games that are violent or feature questionable portrayals may help provide solutions to those problems. Moreover, because the video game industry is seen as a desirable field to be employed in, with various policy and education institutions targeting game development for funding, an understanding of how different the video game industry is from other industries is needed. Changing technologies, including mobile phones and cloud-based services, are forcing the industry to adapt both its products and its structure. Time will tell whether these technologies represent a significant threat of disruption to an industry that is heavily concentrated and focused on only a small portion of the global market.

Contextualising the potential for such a change is particularly important, as the industry, while global in terms of production, is much more limited in terms of consumption. Estimates suggest that eight countries made up approximately 80 per cent of global video game hardware consumption in 2009. The United States, Canada, Japan and a number of countries in Western Europe are the major consumers of video game products. Software production is even more concentrated, with three countries dominating the production of the bestselling video games of all time: the United States, Japan and the United Kingdom (Nichols, 2013). Such concentration has left many countries and regions to fend for themselves. Countries currently representing markets so small that sales data are rarely included in industry reports – such as Brazil, Argentina and India – are all likely to become important sites for game development (Peréz Fernández, 2013; Portnow *et al.*, 2013; Shaw, 2013). In some cases, these differences can be explained in relation to infrastructure, particularly as the video game industry continues to push towards cloud-based gaming, which requires high penetration of networked technology and reliable, high bandwidth (Aslinger, 2010, 2013). Similarly, the advent of mobile gaming and of digital distribution has opened up game development in places poorly served by the mainstream industry (Moss, 2013). In other cases, political and cultural differences may play in. In a number of countries in the Middle East, for example, one limit on the penetration of video games is due to state concern with their apparent Western influences (Lien, 2013; Šisler, 2013). Finally, there are countries that are also limited by simple economics and labour factors. Lebanon, for example, has struggled bringing together workers who have the skills to make games, even as it has struggled with the political dimensions of gaming (Lien, 2013). In contrast, some countries able to supply the raw materials for the industry's products or where they may be assembled include the Democratic Republic of the Congo, Ethiopia, Mozambique, Bolivia, Peru and China though the high cost of the final product leaves many unable to afford them.

THE STUDY OF VIDEO GAMES

The formal study of video games is just beginning. As games have gained in popularity, with expanding audiences and incorporation of more sophisticated technologies, they are now recognised as a unique embodiment of culture worthy of study. It is not uncommon for this study to come under the lens of cultural and textual analysis, research which insists that video games can have legitimate artistic value because they can – though not always – maintain complex narratives and design elements. More than simple entertainment, video games have become texts to be unpacked and analysed.

The ability of video games to fulfil ideological roles has resulted in their reevaluation by policy makers. Governments around the world have begun to ask questions and seek solutions to the problems and potentials raised by video games. Even seemingly unlikely institutions such as the US Army have increasingly integrated video games into their recruiting efforts (AP, 2002; Huntemann, 2009). Beyond this ideological and pedagogical potential, there are other industry-related concerns that, until recently, have received less attention. First, a number of European countries and American states are actively working to subsidise local video game production because it is seen as a fast-growing, highly profitable, competitive industry. Second, questions dealing with software piracy and intellectual property are posing problems to the industry but have received little attention compared to piracy in film and recorded music.

Because of this, it is surprising that the production of video games – the understanding of how the industry is structured and why – has been left largely unaddressed. Since researchers, consumers and policy makers have emphasised that video games can have profound ideological implications, it is ironic that they have largely ignored questions of how and by whom video games are actually created. This lack of attention is, in some part, due to the lack of respect given to the cultural commodity of video games and of the industry that produces them. Long considered a minor sector of other more important industries, it was not difficult to dismiss video games as inconsequential toys for children. It also owes to the success the industry has had at self-description and regulation.

Video games have become more than just a subset of the computer or toy industries and, while they are often produced in conjunction with films and television shows, they are distinct entities. Thought similar to all of those products, video games are produced by an important industry, worthy of consideration on its own merits. Its level of concentration and its relations between labour and management should both be taken into account when thinking of an industry as more or less desirable.

This study examines the mainstream production of video games in order to better understand the industry and what Bernard Miege refers to as the 'logics of production' (Miege, 1989). Figure 1 diagrams the key factors the video game industry has navigated in order to produce the particular logics of production that govern it. The figure, which draws on Sousa's (2010) explanation of how national culture and regulation could result in distinctions within an industry, elaborates on the range of decisions made both at an industrial and product level. What it suggests is that a number of different factors – audiences and how to engage with them, intra- and inter-industry considerations and the role and location of state control on the industry – impact how video game production works and, in turn, the ways in which video game production itself can change how it relates to those factors in the future. In other words, a different set of choices produces a different sort of production and industry. Video game production could work in various ways. This is important because, while this book focuses on mainstream production, there exist both new challenges and a growing resistance to that industry and, indeed, games are produced that attempt to ignore or subvert the logics of the mainstream industry. Choices about what logics to follow help to create the rules of the distinct markets in which video games

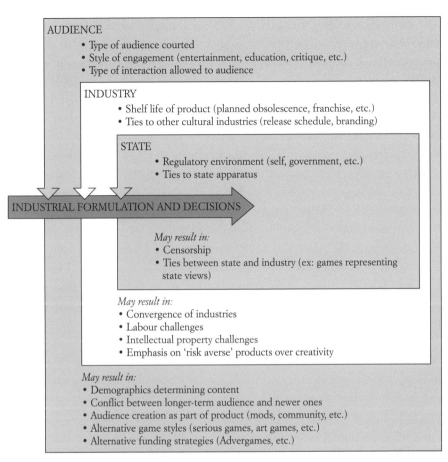

AUDIENCE
- Type of audience courted
- Style of engagement (entertainment, education, critique, etc.)
- Type of interaction allowed to audience

INDUSTRY
- Shelf life of product (planned obsolescence, franchise, etc.)
- Ties to other cultural industries (release schedule, branding)

STATE
- Regulatory environment (self, government, etc.)
- Ties to state apparatus

INDUSTRIAL FORMULATION AND DECISIONS

May result in:
- Censorship
- Ties between state and industry (ex: games representing state views)

May result in:
- Convergence of industries
- Labour challenges
- Intellectual property challenges
- Emphasis on 'risk averse' products over creativity

May result in:
- Demographics determining content
- Conflict between longer-term audience and newer ones
- Audience creation as part of product (mods, community, etc.)
- Alternative game styles (serious games, art games, etc.)
- Alternative funding strategies (Advergames, etc.)

Figure 1 Forces Impacting Video Game Cultural Logics
Source: Based on Sousa (2010).

are made and sold, as well as labour patterns, methods of production, distribution, marketing and retail practices which bring video games to the point of consumption. Such logics suggest not only the ways in which a particular industry or institution works but also the reasons why it must work in these ways. The logics of the production of video game commodities are a key factor in the messages and ideologies conveyed. Central to the understanding of an industry and its logics is an examination of a product's markets and consumers – in this case, audiences. As this study shows, the markets and audiences for video games are more significant and varied demographically, though still highly concentrated geographically and economically, than is often acknowledged.

However, this study does not attempt to address how a video game will affect players or what those players will take from (or bring to) a video game. This book does not seek to explain how messages are constructed or why certain messages are more popular than others, except by suggesting that particular types of messages may be more advantageous economically. Instead, this study seeks to provide at least part of the structure that surrounds these questions. If we want to understand the significance of any game, whether it is *Pac-Man* (1980), *Grand Theft Auto: Vice City* (2002), *Bully* (2006) or *Manhunt* (1983), or even *America's Army*, one of the things we need to know is how making a game serves the interest of its producers. Once we have a sense of how and why they made a game, we can better address questions about the effects and meanings of those games.

Video games are cultural commodities – the products of a cultural industry organised through the capitalist exchange of goods. As with most media commodities, they have the potential for considerable ideological influence. In part, the conditions under which they are produced and the forces to which they must respond impact the ideologies video games convey. In order to better understand these decisions, this study examines not only the industry as a whole but also the individuals and institutions involved in production. It places video games into a historical context and supplies a motivation for the particulars of production.

How video games are produced has helped establish the business as a major media industry in its own right. As with other media industries, the video game industry's status owes to its successful courting of a highly diverse group of consumers. But it also relies on a high degree of concentration, tight control over the products and who can produce them, and on maintaining its control over content rather than allowing outside regulation. These factors have heavily impacted the relationship between labour and management within the industry as well as how video games are tied to other media. This has played out in two major areas: the convergence of technology and the rise of advertising.

VIDEO GAME ECONOMICS
Although limited, there has been some examination of the economics of video games. One of the earliest economic examinations provided an excellent linkage between their production and the Western military-industrial-communication complex (Toles, 1985). Little attention has been given to these ties following Toles's study, but in her work lies the foundation for a critical understanding of the video game industry. Instead, later studies have typically focused on the software side, ignoring the production of the various hardware commodities that are also vital to success.

A second thread of studies has focused on the portrayals of economic systems within video games. A number of games, in particular online ones such as *Ultima Online* (1997) and *EverQuest* (1999), have developed peculiar 'virtual economies', which have begun to spill over into the real world (Castronova,

2001, 2002, 2005). Typically this has occurred as a result of the accumulation of virtual property that is then sold to other players in the real world. The sale of virtual property has become a chief source of revenue for online games and has gained the attention of regulators in countries and regions, including China and South Korea, where online gaming is particularly popular.

Others have moved beyond this to suggest that the study of video games needs to more accurately portray the targeted audiences (Newman, 2002). This will be discussed in more detail in Chapter 2 as part of the analysis of the current industry structure. The audience commodity is crucial to the industry and has fallen, as have most areas of technology, into gendered categorisations (Meehan, 2002). This is a dangerous precedent and must be avoided because, as discussed in Chapter 2, the audience has had to shift in response to evolving logics of consumption.

One move to understand video games as an economic force emerges from the industry and related sectors. These studies recognise the games as entertainment but also as marketing and policy forces that can be understood, controlled and/or exploited. On one hand, these studies seek to deal with the implied danger of marketing violent content to children as in Anders's 1999 study. Ultimately, her examination suggests that the industry has to toe a delicate line on the issue of violence, though she does not address the overemphasis on violence in video games in media and research (Anders, 1999). The relation between video games and marketing has only grown more complex as the industry has matured. The modern industry has developed a complex strategy, which may rely on marketing at different phases of the development process, the use of players' in marketing efforts and the range of ways to use games as marketing devices (Zackariasson and Wilson, 2012).

Thomas Hemphill attempts to address the problem of violence in entertainment industries, including video games, by suggesting alternative issue management techniques and ways entertainment media can take better advantage of political views within the US. While his work does not focus explicitly on video games, his suggestions and concerns are mirrored in the literature – particularly in the news media (Hemphill, 2002).

More recently, however, researchers have begun to acknowledge the similarities and differences between the video game industry and other media forms. Dmitri Williams lays out the initial structure of the industry, consisting of publishers, developers and hardware manufacturers. He also acknowledges the role of retail and advertising in the economics of the industry. Williams's work, however, exhibits little concern with the level of concentration in the industry. Moreover, he does not address the problematic labour situation or any of the trickier matters of property control (Williams, 2002, 2003a).

In contrast, Nick Dyer-Witheford offers a critical view of the labour practices, noting that the industry is increasingly taking advantage of transnational labour. He also points out that this labour is highly gendered (Dyer-Witheford, 1999). In later work, he begins to suggest a class structure within the industry that tends to fall along income and educational lines in addition to those of gender and nationality (Dyer-Witheford, 2002; Dyer-Witheford and De Peuter, 2009). Like much of the study of game hardware production, however, these studies focus on console production ignoring, in particular, the impact of handheld games on the industry. One of the best examinations of the global dynamics at play within video game production, his work suggests many questions about the nature of industrialised video game labour.

As Zackariasson and Wilson note in relation to marketing, game players often contribute more than their consumption (2012). The industry also exhibits a particular set of labour practices that rely on the work of consumers. Most notable of these is the concept of 'modding' or the creation of game expansions by fans, which tends to occur particularly in the action and Role-Playing Game (RPG) genres. These, however, are only some of the ways players might take on the peculiar function of labour in the industry (Postigo, 2003).

Along similar lines, Klang examines the struggles of ownership between fans and the industry. Drawing on ideas raised by Castronova, Klang examines the ways in which 'avatars' – or the representations of players in games, particularly RPG and Massive Multiplayer Online Role-Playing Games (MMORPG) – become contested zones of intellectual property (Klang, 2004). Raising more questions than it answers, Klang's study suggests a number of interesting directions in which video games may force the ongoing intellectual property debates plaguing communication industries. These questions are particularly crucial in China and other countries where online gaming is popular (Chung and Fung, 2013; Ihlwan, 2007; Rosenberg, 2010).

Two of the most in-depth examinations of the industry have come from European scholars. The first focuses on the battle in the European market between 16-bit consoles in the mid-1990s (Hayes et al., 1995). While primarily a historical study, a number of illuminating features are first analysed here in the economic literature. Europe has historically been treated as a secondary market for video games, with hardware lagging roughly a generation behind. As will be discussed in Chapter 1, one measure of the industry's history is marked by console generation. Each new issue of a console, which tends to happen every three to five years, is marked as a new generation. For much of the industry's history, the consoles available in many markets, including Europe, were one generation behind what was available in the North American and Japanese markets. It is in the period scrutinised by Hayes et al. (1995) that

Europe begins to emerge as a primary market with its own systems of production and distribution. But the authors also discuss the use of planned obsolescence as a marketing tool, something acknowledged in earlier (primarily industry) literature, but not examined.

In contrast, Kerr (2006a) details a more recent and more global perspective on the video game industry. Kerr situates video game production firmly within the question of video game meaning, allowing the global nature of the industry to be seen as a factor in the ways in which games themselves are used. This also reinforces the view that video games must be seen as creations of culture even as they are tools that can enforce a culture's dominant views. Kerr also furnishes a useful sketch of the European market in spite of the scarce data available, providing a good sense of not only how games are made but of how the costs for them break down in the European case (Kerr, 2006a). She has also examined the challenges the UK and Irish game industries face in a globalised production system, with increasing labour shortages (Kerr, 2012). The questions her work raises are particularly important as the video game industry has become a focus of a broad range of policy initiatives within the European Union and a number of individual member nations (Kerr, 2013).

The book *Digital Play* examines similar issues, drawing on political economic concerns as a means of discussing the marketing of video games as a cultural force for audiences to respond to (Kline *et al.*, 2003). While proffering some excellent frameworks for understanding the overall layout of the industry, little attention is given here to the system of production itself. Rather, video games are treated as cultural texts that must be understood in terms of their messages and marketing as a system for audience response and understanding.

Finally, the nature of competition internationally within the industry has raised the question of territorial lockouts. Territorial lockouts occur when the industry creates products that only work in particular geographic markets – for example, DVDs that will play on machines sold for a specific market. The video game industry also creates products with these lockouts in mind. Ip and Jacobs (2004) attempt to examine reasons for this practice, concluding that the video game industry seems unsure about it, but has tended to follow it for almost traditional reasons: to protect against piracy and to spur creativity. However, their study suggests that both consumers and producers are increasingly sceptical about the practice. Such a practice also raises important concerns in regards to cultural imperialism that go unaddressed within the study and the industry.

These studies often suggest that it is important to think of video games as more than an American or Japanese product. Like the film and larger computer industry, the video game industry has gone global, incorporating not only global audiences but global forms of production as well (Kerr and Flynn,

2002). In addition, the industry has cemented ties and borrowed forms from other industries, most notably film (Howells, 2002). This trend has continued and is reflected in the current structure of the industry discussed in Chapter 4.

The field of video game studies has grown considerably in recent years. It is imperative that this field pay attention to the institutional nature of the commodities and texts being studied in order to better ground their understanding of how they are used and what they mean (Douglas, 2002). Video game commodities have evolved beyond being toys for adolescent boys; indeed, they were never that simple. Video games must be viewed as unique cultural artifacts – both tool and art – that can be used in a variety of ways for a variety of purposes by a range of players. Video games are capable of both reinforcing and subverting ideologies and stereotypes. However, what the field of video game studies is lacking must also be considered.

It seems clear that to better make sense of how video games work in terms of play and meaning, an in-depth, systematic analysis of the industry within its historical context is necessary. Based on the existing literature discussed above, a number of important trends warrant examination at the institutional level. First, the video game industry has become one of the dominant cultural industries of our age that earns billions in revenue; employs a substantial, globalised labour force; and draws the attention of a sizeable audience across a range of platforms. Second, the industry is tied to a number of other industries including the film and computer industries, becoming an important factor in their profitability. It has also attracted increasing attention from policy makers because, as a burgeoning sector in the information and creative industries, it is seen as important to growth and development. Third, the continued anxiety about the effects of video games requires a consideration of how games are produced. Despite some public concern – and even state interest – about video game effects, the policing of content has all too often been left to the industry. Fourth, the rise of video game studies as a field of academic interest has resulted in the creation of game studies programmes at a number of universities around the world. These programmes have focused largely on professional development. If only for the pragmatic reason of understanding the industries such programmes are training students for, a systematic study of the video game industry is required. Fifth, histories of the industry have managed to present it in a largely ahistorical and acultural fashion. Some attempt that considers the industry in relation to the events, institutions and culture surrounding its development is necessary. Finally, the shift felt in both effects and cultural research regarding who uses video games and to what end calls for a more thorough understanding of the practices engaged in directing the industry's activities: at whom are video games targeted and how does the industry target them? The following chapters pick up on these themes, and elaborate on them, focusing on the mainstream industry and its development.

Chapter 1 focuses on the historical emergence of the video game industry. It looks at historical trends related to hardware production, software creation and audiences that influenced the industry's emergence. It also provides case studies of three companies – Atari, Nintendo and Sega – which each influenced the shape of the modern industry. Chapter 2 begins the examination of the modern video game industry's structure and considers the changing nature of the video game audience. It then focuses on the software side of the industry. It includes a case study of Electronic Arts and Activision-Blizzard in order to tease out distinctions between software publishing and development. It also looks at the importance of software genres as a production category and discusses some of the regulatory challenges for video game software. Chapter 3 turns to the other side of the house: hardware production and the process of distribution, whether digitally or via arcades or retail stores. Case studies of Microsoft and Sony raise questions about the global nature of hardware production as well as the challenges posed by advances in digital distribution. Chapter 4 focuses on the relationship between the video game industry and other cultural industries. It analyses the industry's impact on the microchip industry. The range of ways in which the industry has made use of licensing and franchises is addressed, as are ties between video games and the film, television, recorded music, publishing, sports and advertising industries. Chapter 5 focuses on labour and production, drawing comparisons between the computer and information industries. It also examines questions of employment within the industry, the importance of players as labourers, and of educational programmes targeting the industry. Case studies of the labour problems faced by Electronic Arts and Rockstar games and of the troubled relationship between the video game industry and Hollywood labour unions raise questions about the future direction of the industry. Finally, the Conclusion draws together the important threads of the book, to suggest questions the industry faces and challenges for video game researchers in the future.

1

An Industrial History of Video Games

While there have been numerous histories of video games, very few have attempted to address the economic context of game production. In order to truly understand video games and their impact on society, it is necessary to understand the industrial system of production from which they emerge. By contextualising them as commodities, it becomes possible to see how clearly video games are tied to a variety of social trends as well as to other cultural industries. By moving beyond histories that focus on when games were produced or which designers were involved in their production, the market forces which shape the games we enjoy – those forces that serve to constrain and direct game development – can be understood.

From almost their very beginning, video games were produced as commodities and were subject to the logics of cultural commodities. As such, games must fit into the larger capitalist framework with content that is likely to sell to the largest possible audience while supporting the existing social structure. The history presented in this chapter attempts to move beyond 'great man' histories, beyond tales of emerging genres and compelling games, and beyond mere technological capability. Instead, it focuses on three aspects the industry must contend with from a production standpoint: institutional and industrial formation, technological reliance and the nature of consumers for its products. This is not to suggest that other themes in the history of video games are unimportant or invalid. Rather, it suggests that we can gain a better understanding of how the modern, mainstream industry and its commodities were achieved by examining these three areas.

By examining video games from this perspective, three major discrepancies become apparent, each of which flies in the face of typical 'myths' about video games and the industry which creates them. The first of these deals with the industry's view of itself in regard to function and similarity to other industries. Many studies, particularly those that look at video games as commodities, consider them a subset of the toy industry (Business Week, 1979a; Elkin, 2003, 2005; Kang, 2005; Vogel, 2004). In fact, the industry really emerged – and has consistently seen itself – as a subset of the computer industry but with distribution models based on film, toys and recorded music.

The second major discrepancy in the existing histories of video games is tied to the first. Any portrayal of video games that views them as toys allows the product to be marginalised into children's culture. While their impact on children is clearly of great importance, we must be careful not to fool ourselves that they are designed to be used only by children. Nor, as the history in this study shows, have they ever been. Particularly in the industry's early history, we see that video games are best suited to an adult audience.

Finally, because the industry and its products have been treated as toys, their capabilities as communication devices have been largely ignored. Producers – particularly console manufacturers – have long recognised these possibilities, however, and this has shaped their goals for what video games can be. Because these capabilities enable video games to become media hubs, allowing users to play games, download and watch movies, store data and even function as telephones, ignoring this potential both misrepresents the industry and underestimates its scope. It also provides an additional reason why a number of companies and industries have repeatedly found themselves drawn to the video game industry. Ignoring video games as devices created to communicate slowed the examination of video games in terms of convergence and ideology. But again, a close examination of the discourse surrounding their development makes clear that video games have always been worthy of consideration in terms of communication and its processes because the convergences currently being witnessed are neither new, incidental nor accidental. The ties between video games and other forms of media are more fully explored in coming chapters, but the evidence of these ties is present from the beginning as is the goal of convergence.

To understand the technological history of video games requires an appreciation of the capabilities of hardware and of where software itself is stored. Additionally, as consoles become more complex, it is worth considering the other functions of video game technology. The history of video games has been typically divided into overly general eras (for example, arcade versus home play or early versus modern) or into overly specific sets of generations that rely on the computing power of the chips involved (for example, the 8 bit generation versus the 32 bit generation). However, neither of these views provides useful detail about video games as communication devices. However, as Table 1.1 shows, taking a more expansive view of the capabilities of the technologies involved we can think about video games as having gone through four distinct epochs. The first epoch, the Foundational Epoch, began in roughly 1972 and was defined by software that was inseparable from the hardware it was played on (Ferris, 1977). Thus, one machine could only play a single game. A common example would be an arcade game, the dominant form of video games in their early history. By

Approximate year	Description	Challenges	Examples
1972	**The Foundational Epoch** • Games and consoles are linked	Finite amount of information can be encoded, limiting use	*Pong* (1972) Arcade games
1976	**The Modularity Epoch** • Games and hardware separate	Modularity allows expansion via new software, but with potential loss of proprietary control	Atari 2600 (1976) Nintendo NES (1985)
2000	**The Convergence Epoch** • Games and hardware separate • Ability to use other media	Potential loss of audience to other media forms and increased attention (whether competition or intervention) from rival media industries	Sony PS2 (2000) Xbox (2001)
2010	**The Networked Epoch** • Games and hardware separate • Ability to use other media • Ability to make other media? • Ability to download content?	Potential competition not tied to hardware, but reliance on high bandwidth might limit audience. Also increased danger of malware (viruses, etc.)	Xbox 360 (2005) Wii (2006) Sony PS3 (2006)

Table 1.1 Historical Epochs in Video Game Technology
Sources: Campbell-Kelly (2003); DeMaria (2003); Ip and Jacobs (2004); Kent (2001).

roughly 1976, however, the Modularity Epoch began, defined by the separation of software from hardware, allowing one system to play multiple games. This change makes the development of independent software possible. Owing to advancements in computer hardware, this generation of video games emerged very quickly and, ultimately, overtook the first generation. This second epoch also ran considerably longer, seeing the industry's structure formalise to help establish means of proprietary control. It is in this period that the division of software labour between developers, who create software, and publishers, who guaranteed quality and handled the distribution of software to platforms and retailers first emerges. By 2000, however, a new epoch – the Convergence Epoch began. In this period, video game platforms evolved into more than just tools for games. They gained the ability to play other media such as DVDs, CDs and to display photos and other user-generated content. This shift made the video game industry an important competitor not just with film and television but also with other forms of distribution, including telecommunication. This period also marks the crucial change of the audience from consumers to consumers and labourers. Finally, around 2010, a fourth epoch, the Networked Epoch, emerged, defined by its adoption of networked logics, allowing not only online cooperative play, but the ability to share information between consoles and to download software rather than purchase physical copies. This shift is not without its challenges, as it requires high bandwidth but also potentially destabilises any dominance within the hardware sector of the industry. The implication of this for distribution will be discussed in Chapter 3, showing its profound impact on the modern video game industry.

This chapter intends to provide an early industrial history of video games, based on that model of historical epochs. It begins by examining the question of what made for the first video game and how that led to the creation of the industry. Next, it examines the first two decades of that industry. While the industry faced a number of crucial challenges, none of the responses was technologically or industrially inevitable. The choices made resulted in the mainstreaming of video games themselves and the evolution of the modern industry. Next, this chapter discusses three key areas of development: the video game hardware segment, the software segment and the growth of the video game consumer base. It concludes with case studies of important historical players that faced key struggles in the early days and that were significant in the development of the modern industry's structure.

SPACEWAR! (1961): THE EMERGENCE OF VIDEO GAMES

As Toles has noted, video games emerged from generously funded government research in communications and computers during the early 1950s and 1960s. The first games were developed as part of the mammoth Advanced Research Projects Agency (ARPA), which resulted in the creation of the Internet, as well as many of the home computer devices which have since become commonplace (Toles, 1985). In some cases, game development was actively supported, as with simulators. However, a considerable number of games were also developed that served no official purpose and received no official funding. Many of the earliest programmers experimented with games as a means of testing the capabilities of the computers they worked with as well as to demonstrate those capabilities to others.

These first games generally dealt with motion, strategy and decision making. In fact, almost since the creation of computers, simple strategy games including Tic-Tac-Toe had been devised. Eventually, more sophisticated games, which pushed the capabilities of both programming and machines, were developed. In this way, games often spurred both programming and technological advances (Becker, 1976). Because of the close ties between initial game development and the US military-industrial complex, it is not surprising that what is often cited as the first video game was a game about spaceships shooting each other. The game, titled *Spacewar!*, was developed in 1961 (Herz, 1997). It is interesting to note that in the decades since, video games have become so legitimised that there seems to be a competition among US government agencies to claim a hand in their initial development (DOE, 2003). While there are a number of contenders for the distinction of the title of the first video game – including *Tennis for Two* (1958), developed by an engineer at the Brookhaven Institute to demonstrate the motion capabilities of oscilloscopes, and *Spacewar!*, which was developed by a programmer working with ARPA at the MIT computer node to demonstrate how computers could display motion (Herz, 1997), this study marks the beginning with *Spacewar!* Unlike *Tennis for Two*, which relied on entirely solid state technology, *Spacewar!* integrated both a video display and computer programming (DeMaria, 2003). It is worth noting that in these earliest games, the intended audience was almost always adults. Regardless of which game one uses as the start, the impact of video games is, as Nick Dyer-Witheford has noted, a watershed moment that warrants comparison to the emergence of other important media such as film and recorded music (Dyer-Witheford, 2002).

While *Spacewar!*'s claim to be the first game is largely technological, it is worth expanding on. Computers had been in use for almost two decades by the time *Spacewar!* was invented and, as noted above, simple strategy games like Tic-Tac-Toe had been created to test computers in their earliest days. But *Spacewar!* redefined what a game using computers could do. Its reliance on a video display alone makes it noteworthy, but it was also one of the first programs that relied on keyboard instructions rather than punch cards. Both features introduced levels of interactivity as well as the possibility for mass-market application. Now, a user did not need to know how to program in order to play. Instead, access and instruction were all that was required. While the game was not initially created for a consumer market, this change made the game and others adaptable for such a market.

The debate over what made for the first video game also marks a rhetorical schism in the study of games: the question of terminology. *Spacewar!*'s development marks the first moment when the term 'video game' could be distinguished from the term 'computer game' or 'electronic game'. Prior to *Spacewar!*, games could be created on a computer that included no screen interface. These are best referred to as computer games. In such cases, one might program via punch card and receive results via printout. Later, the toy industry began to market games that relied on microprocessors and computerised parts but that included no video interface. Games such as the 1978 release *Simon* or many modern pinball games are examples. These are best referred to as electronic games. Lastly, there are products that combine both computerised parts and a screen interface. These we will refer to as video games.

The period in which *Spacewar!* and *Tennis for Two* were developed functions as a sort of pre-industrial phase for video games. In this period, game development logics focused on different concerns than those of either the early industry discussed later in this chapter or the modern industry scrutinised in the following chapters. In this period, games were designed with a focus on testing the limits of technology and of pushing the advancement of both programming and technology. While they often had an element of entertainment to them, they were also learning tools and objects of enculturation.

Spacewar! is also significant because of how rapidly it spread. Its creator, Steve Russell, admits to having considered whether he could market the game, but he ultimately decided it was too difficult to reach the audience for them at that time (Ferris, 1977). Thus, from their earliest days, video games were seen as a potentially lucrative commodity; that it took almost a decade for capitalist forces to recognise the potential only suggests there may have been difficulty in finding a market or making the technology affordable. Part of the question of affordability is down to the lack of industrial support for cheap production. At the time of *Spacewar!*'s creation, computers were manufactured individually, and the components were both too large and too expensive for general use. It would take almost a decade from the game's creation for cheap manufacturing to catch up, making consumer-oriented video games a possibility.

Instead, Russell allowed the source code to be shared across the ARPA network to anyone who was interested. The game took off across the ARPA nodes, with programmers adding refinements and new features. The uptake of the game was so quick and widespread that by the mid-1960s it was estimated that there was a copy of *Spacewar!* on every computer in the US (Herz, 1997). By the early 1970s, *Spacewar!* was being played nightly, during the typical downtime for mainframes, on computers at many major companies as well (Ferris, 1977). Interestingly, *Spacewar!* also served to legitimise computer science to a more general public (Herz, 1997). Russell notes that many programmers would show *Spacewar!* to friends and family as a means of explaining not only how computers worked but also what they could potentially do. Other games developed as well, and the most popular on mainframes after *Spacewar!* was one based on the 1960s television show *Star Trek* (Ferris, 1977).

EMERGENCE OF THE INDUSTRY

[Home Computers] are the railroad train of the 80s.

Robert F. Wickham (Business Week, 1979a)

While its creator was unsure of the market for video games, *Spacewar!* laid the groundwork for the industry to emerge. It provided an initial audience, a means of legitimisation and an entertaining use for a technology that had previously been unavailable to most people. Coupled with the rapid development in both computer hardware and software, commoditised video games became a legitimate possibility. The rapid increases in computer memory capability, which Moore's Law predicts as doubling approximately every eighteen months, removed one of the biggest limits on computer – and video game – development. Russell's game, which demonstrated that a game could succeed without requiring players with programming knowledge, combined with the availability of increasingly powerful microchips, set the stage for the emergence of the video game as a commodity.

With these key questions addressed, a consumer-oriented video game industry was viable. Within a decade of *Spacewar!*, the first successful video game company was founded by Nolan Bushnell (Campbell-Kelly, 2003). Atari was initially formed to market the arcade game *Pong* (1972), through alliances with established toy merchandisers (Business Week, 1976a; Cohen, 1984). In very short order, the company moved to producing home consoles and software. While it is perhaps best known for its creation of gaming consoles in the 1970s, Atari is still an important force in the industry today. Bushnell first pitched his video games to arcade producers and, while the games were successful, he received little of the profits due to licensing deals he signed for its initial production as well as a later lawsuit by Magnavox over patent infringement. This prompted him to form his own company (Cohen, 1984). The company began with the production of coin-operated arcade games. At this time, computer technology was neither sufficiently accepted nor accessible to the general public, forcing the first industry developments to focus on arcade games.

At the same time that Atari was forming, major toy companies were the primary provider of electronic games such as *Simon*, various electronic sports games and others (Kent, 2001; Rice, 1979; Salsberg, 1977). Among the firms manufacturing these games were Milton Bradley, Fairchild Electronics and RCA (Business Week, 1979b; McQuade, 1979; Salsberg, 1977). These companies also experimented with creating video games, but focused on first-generation games (DeMaria, 2003). Their size relative to the newly emerging Atari allowed these companies to define the market's production schedules, centring around the Christmas shopping season. The fact that the major toy and electronic game manufacturers concentrated on arcade games allowed smaller concerns to spring up and redefine the game industry by focusing on the possibility of the home as the site of play (McQuade, 1979). To a large extent, the success of Atari dictated the fortunes of the video game industry overall. The company had tremendous success into the early 1980s, before a series of crucial errors, discussed shortly, sent both the industry and the company into decline. It was not until the rise of Nintendo and Sega in the late 1980s and early 1990s that the video game industry rebounded. Similarly, the arcade industry peaked around 1982, with around $2 billion in revenue (Campbell-Kelly, 2003). The manufacture of arcade games is still big business, serving as test markets for upcoming console and PC games (McCallister, 2005). The revenue it generates for the game industry, however, has become much less crucial. With this in mind, the remainder of this chapter focuses on the development of video game hardware sectors of personal computers, consoles and handhelds.

In contrast to these more established companies, Atari's development in many ways served as a precursor to the home computer industry, which would begin to emerge only a few short years after Atari's founding and owes a considerable debt to the company's first forays into arcade production (Campbell-Kelly, 2003). By 1975, game production of second-generation systems was possible, with the creation of consoles which played games on cartridges. This in turn allowed game systems more flexibility, making them more attractive to consumers. Atari, however, was the first attempt by the industry to reconcile the existing logics of the toy industry and the computer industry. From the toy business, video game manufacturers adopted a production scheduled centred on sales around the holidays, and from the computer industry, they adapted to the advancements in processing power to rely on planned obsolescence for their products. Atari's rise marks the shift from the Foundational Epoch to the Modularity Epoch; its first game, *Pong*, is perhaps the best example of the former, while the Atari 2600 console is one of the earliest and most successful examples of the latter. The Atari 2600 resulted in the rise of independent game developers, piggybacking on its success. The company's lack of control of those developers proved one of the biggest hurdles, and threatened to destroy the industry. Eventually, as the industry worked towards the Convergence Epoch, it would also adopt some of the logics of other media industries, particularly the use of tighter intellectual property controls and structure. The example of Nintendo later in this chapter examines that shift.

CASE STUDY: ATARI

The example of Atari is an instructive one in the early history of the video game industry, illustrating its volatility prior to stabilising in the early 1990s. It

Year formed	1983 (original company, 1972)
Headquarters	Paris, France
2012 Sales (US$m)	53.2
2012 Employees	354
Industry sector	**Software publisher**
Key software franchises	*Dragon Ball Z*
	The Matrix
	Unreal Tournament
	Neverwinter Nights
	Atari Flashback
	Dungeons and Dragons
	Roller Coaster Tycoon
	Test Drive
	Godzilla
	Act of War
	Pong
	Asteroids
	Centipede
Studios owned	Atari London
	Atari Interactive
	Atari, Inc.
	Eden Games
	Humongous Games

Table 1.2 Corporate Profile of Atari, S.A., 2012
Sources: Hoover's (2012); Mergent (2008).

demonstrates a number of the key problems hampering advancement. Table 1.2 provides a breakdown of Atari's modern status, including key franchises and a brief corporate history.

In addition, Atari represents one of the first examples of the modern industry structure, in which a hardware maker also began to publish software and seek out deals with smaller developers. This gave Atari immense power in the market and, when the company struggled, this was felt throughout the industry as a whole. In particular, the relations between Atari and smaller developers suggested some of the battles over intellectual property control that later companies, such as Nintendo, and the modern industry would face.

Founded in 1972 by Nolan Bushnell in Sunnyvale, California, Atari rode the wave of the industry from arcades to consoles, from hardware production to software production, and from outside interest. As an early Silicon Valley firm, Atari drew heavily on programmers and designers from universities affiliated with the ARPA project. Were it not for the struggles for stability Atari navigated before many other companies, it would probably be a bigger power in the modern industry. Founded for only $250, Atari watched its sales grow to $3.2 million in 1973 and to $39 million by 1975. The company's first success was the arcade hit *Pong* (Cohen, 1984). Demand was high, but Atari failed to gather enough capital to meet production demands, losing considerable revenue. Bushnell explained: 'it takes a lot of cash to build a $20 million industry for a three-month selling season' (Business Week, 1976a). The problem of capital was an ongoing one for Atari as was the reliance on the toy industry's model of seasonal sales. In 1974, an attempted follow-up to *Pong* was plagued by bugs, costing the company $500,000 – as much as its revenue in the previous year.

Almost as quickly as it had arrived, Atari faded into the background. In 1976, Bushnell sold Atari to Warner Communications. For Warner, the buy was sensible: Atari had sold another $39 million in goods and earned a $3.9 million profit. Estimates suggested that it could sell as many as 500 million units by 1980. If that were the case, the company stood to gain $500 million in sales (Business Week, 1976a). This was only briefly the case. In 1981, there was a video game machine in 17 per cent of American households, a growth of almost 8 per cent from the previous year (Cohen, 1984). Atari released a string of hit console games including *Pac-Man*, *Space Invaders* (1978) and *Asteroids* (1979) (Kent, 2001). *Pac-Man* alone sold more than 2 million units for the company in 1981, becoming one of the first and most impressive cultural events in the video game industry, spawning songs, merchandise, spin-off games and even a Saturday morning cartoon show (Cohen, 1984). Of those games, however, only *Asteroids* was developed in-house; Atari had switched to licensing games from outside and began reusing existing game engines to produce games in-house. Because of the relative simplicity of the machines themselves and of the programming models, it was easy for small software developers to create cheaper games for established systems. Figure 1.1 teases out both the historical sectors of the industry structure at the time of Atari's dominance and how those sectors related to each other via production logics. Despite heavy concentration in the market, the relative ease of entry by software developers kept the industry competitive and limited the dominance of hardware manufacturers. Only when problems emerged with a system of unchecked developers creating poor-quality games that were difficult to distinguish in the market would the industry structure shift.

By 1982, the fact that Atari controlled 80 per cent of the video game console market in the US would seem to have made its position unassailable. Moreover, the company had just signed the first deal with Hollywood in the industry's history, allowing it to make video games for the hit movies *E.T. the Extra-Terrestrial* (1982) and *Raiders of the Lost Ark* (1981) (Cohen, 2005).

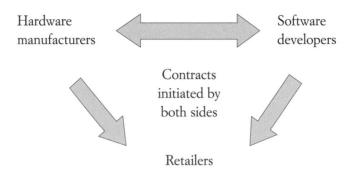

Figure 1.1 Early Market Structure of the Video Game Industry

Licensing of properties from Hollywood would become a profitable strategy for the industry and would eventually suggest new ways for it to develop (Nichols, 2008).

The rights for *E.T.* were licensed from Steven Spielberg in July 1982, just after the movie's release. Atari had promised to have games on the shelves in time for Christmas – in some estimates, even as early as September – a move that left little time for testing and marketing. Atari also ordered 5 million units of the game cartridges, anticipating an easy success (Kent, 2001). Because the production time was so short, the company decided to modify an existing game, spending little time developing any plot, meaning that the game was a dramatic failure. Nearly all of the cartridges manufactured were returned.

This left Atari in a delicate spot. Nearly all of its focus for 1982 had been pinned to the success of the game. When it failed, developers responded by raising licensing costs for Atari, forcing it to pay exorbitant rates even as its sales were plummeting (Kent, 2001). This prompted Warner to rethink its venture, and in early 1983, the company was again on the auction block (Nichols, 1988).

Atari's next owner, Jack Tramiel, was better known for his success with Toronto-based computer hardware firm Commodore International Ltd (Nichols, 1988). Tramiel, who had recently been ousted from Commodore, hoped to restore Atari to greatness by using the same tactics he had at his previous company. Commodore, once the leader in the personal computer industry, had made its fortune by cutting costs below competitors until they couldn't afford to stay in the market (Kafner, 1986). Tramiel purchased Atari for $100 million of his own money in addition to assuming more than $300 million in promissory notes. One of Tramiel's first moves was to extend Atari's reach beyond just video games and consoles further into the personal computer area (Nichols, 1988). The company's computers did very well in Europe, which was then a secondary market for both the computer and video game industries, but it never did well in the US (Shao, 1988). Though focusing on personal computers, it continued to work with video games as well, though the industry in the mid-1980s was struggling and had been largely dismissed as a fad whose time had passed. Tramiel himself recognised the importance of video games, telling naysayers years before Nintendo would revitalise the industry, 'There will be peaks and valleys, but the category [of video games] will never die' (Shao, 1988).

Tramiel and Atari had some obstacles to overcome, not the least of which was Tramiel himself. As the leader of Commodore, Tramiel upset retailers in the industry – at the time, primarily small, independent stores selling nothing but computer games – by bypassing them in favour of larger concerns where he could put his company's products on the shelves at a lower cost (Wise, 1985). At the time, this was a risky move as most computer sales occurred in smaller stores. However, as will be discussed in Chapter 3, today most computer and video games are currently sold in the larger chains, marking Tramiel's move as a visionary one.

Atari also faced two distinct economic challenges: trying to gain ground in the already well-defined and controlled computer market and cutting the huge accrued debt. Tramiel's first action was to institute massive layoffs, resulting in a streamlining of the company's production processes and its books (Nichols, 1988; Wise, 1985). Tramiel's solution was a surprising one for the time: he invested more heavily in the European market (Shao, 1988). Atari's personal computers flourished, becoming the fifth bestselling brand in Europe. This step presaged Nintendo's later shift into the European market, greatly expanding the video game industry's profitability.

Tramiel's Atari was innovative on the video game front as well. First, he re-signed company founder Nolan Bushnell to a $5 million contract to produce games (Shao, 1988). He also began advertising. In 1988, Atari doubled its advertising budget to $10 million, focusing on the video game portions of its business. Finally, Tramiel bucked the system of publisher/developer control exemplified by growing industry giant, Nintendo. As Nintendo pushed into the industry, it did so through a rigid system of control: in order to develop games for Nintendo's systems, developers had to agree to give Nintendo authorisation to release the game. Tramiel took the view that this system was bad for business, and he had Atari produce the first game for a Nintendo system, the NES, which wasn't specifically authorised by Nintendo. Ultimately, this resulted in a series of lawsuits, but it also re-established the idea of independent software development in an industry that had become too guarded following the failure of E.T. and other products earlier in the decade (Newsweek, 1989). Ultimately, most major hardware manufacturers would become software publishers as well, but the lawsuit helped to stabilise the industry by limiting the control of publishers while still allowing them to ensure quality products.

Atari's success under Tramiel is noteworthy for a few reasons. First, the company was profitable within two years of his takeover (Kafner, 1986). Under his leadership not only were ties between video games and the computer industry cemented, but also the retail system between game publishers and retailers took on the form it has today. Finally, it was under Tramiel's leadership that Atari reignited the idea of independent game development, weakening slightly the control of publishers – something another company, Electronic Arts, was to take tremendous advantage of by the mid-1990s.

As the leader of the software sector in the modern industry, Electronic Arts (EA) will be discussed at greater length in Chapter 2. However, EA's origins trace back to the 1980s. Originally, the company produced not just video games but also productivity and educational software, primarily for the growing personal computer market (People, 1985). By the 1990s, the company had recognised the advantages in focusing on games and so shifted its efforts to concentrate on their production. EA modelled itself on the film industry, which had flirted with video games, beginning with Warner's acquisition of Atari (Brandt, 1987; Nichols, 2008). Rather than seek distributors, EA began to forge ties with retailers to secure shelf space in the mid-1980s as a way to insulate itself from big changes in the market (People, 1985). This early move enabled the company to make the leap from developer to publisher within a decade, overtaking other firms, including Atari. EA's founder, Trip Hawkins, is said to have formed EA with an eye on the Hollywood studio system, intending that his company would become the equivalent of a major studio, determining which games – including those developed by smaller studios – would ultimately go on the market (Pitta, 1990).

By the early 1990s, Atari was doing well, though most of its profits had come from lawsuits rather than sales, particularly its antitrust lawsuit against Nintendo. The video game industry was also rebounding, buoyed by the successes of Nintendo and Sega (Wapshott, 1999). Atari repositioned itself to enter the growing handheld console market with the Lynx, but product shortages saw the company in trouble again (Harmon, 1994; Thompson, 1994). These setbacks left the company in a rough spot, and it was quickly bought up by JTS Corp in 1996 and then resold in 1998 to Hasbro and then again in 2000, when French software publisher Infogames, initiated a series of acquisitions and sales involving the Atari properties (Atari, 2008).

Though it was the first major video game publisher, Atari has gone through a number of ups and downs in its thirty-six-year history. In 2008, Atari's acquisition and relaunch as a publicly traded company was completed, with Infogames Entertainment maintaining a majority stake to help take advantage of the Atari name after slumping sales (Hoover's, 2008). Atari, Inc. became the firm's US face. The company has produced and distributed games for all major console platforms as well as for personal computers (Atari, 2008). Starting in 2005, however, the company faced financial difficulties and scaled back its own developments, focusing on publishing software from outside developers. This also prompted the sale of all of its own studios, including Reflections and Shiny Studios, though Infogames still maintains its own development studios (Atari, 2008). Through its board of directors, the company has ties to Delta Airlines, Silicone Graphics, Inc. and music giant Sony BMG (Atari, 2008).

Atari's attempt at the Lynx marks a significant moment in video game history: the move from home consoles to portable handhelds. As will be seen when Nintendo is discussed, the handheld video game represented a significant revitalisation for the industry, in part because it created both a new product market and a new audience: children. What should also be clear from this early history of the industry is that it was largely – and often problematically – contingent upon the factors of production borrowed from both the toy and the computer industries. A number of trends established in the early years are still evident today, though they have been tempered by closer adherence to film and recorded music production norms. The first of the early trends was the focus on fourth-quarter sales – the Christmas buying season – for its products. From the computer industry, it adopted ideas of planned obsolescence, particularly in regard to the continued development of microchips. From film and recorded music, it would adopt means of controlling content, licensing and loss-leader manufacturing.

With this in mind, an examination of the two primary commodity sectors – hardware and software – is possible. These two sectors reflect the impact of decline in video game production. Initially, the companies involved typically focused on the production of one of the two commodity sectors. Ultimately, this would allow outside media companies to take a more active role in the industry.

HISTORIC TRENDS
Hardware

> [Video games] are like the record business. You don't put two and two together and say you're going to sell this many. It's entertainment.
>
> Allan Alcorn, former vice president for Research and
> Development at Atari (Fincher, 1978)

Interwoven with the history of the institutions of the video game industry has been the history of computer processors. Just as Steve Russell's *Spacewar!* could not be bought and sold profitably because computers in the US were so costly, often requiring grants from the government, the emerging industry had to deal with how to make products small and cheap enough to be affordable to the average consumer. The earliest video games could only be placed in arcades and bars because the hardware was so large (Bolter and Grusin, 2000). Similarly, the graphical sophistication in the earliest video games was limited because of the expense of manufacturing systems capable of anything more complicated. As miniaturisation of chips progressed and microprocessors became more readily available, the profit in the industry shifted from arcades to home systems. As chips became smaller and smaller, the hardware types, or platforms, on which a game could be played continued to evolve as well. Table 1.3 details the hardware consoles available by decade. What is clear from the table is both the changing nature of the companies involved, with a variety of electronics and toy firms playing prominent roles in the 1970s and 1980s changing to a domination by a smaller number of video game companies starting in the mid-1980s to today. An increasing concentration in the hardware industry is also apparent.

1970–80	
Atari PONG	Atari VCS 2600
Magnavox Odyssey	Atari Video Pinball
Magnavox Odyssey 100	Coleco Telstar Alpha
Magnavox Odyssey 200	Coleco Telstar Combat
Atari Super PONG	Magnavox Odyssey 2000
Coleco Telstar	Magnavox Odyssey 3000
Coleco Telstar Classic	Magnavox Odyssey 4000
Fairchild Channel F	Atari 400
Magnavox Odyssey 300	Bally Professional Arcade
Magnavox Odyssey 400	Coleco Telstar Arcade
Magnavox Odyssey 500	Coleco Telstar Colortron
RCA Studio II	Coleco Telstar Gemini
Wonder Wizard 1702	Magnavox Odyssey
Milton Bradley Microvision	Mattel Intellivision
Atari Stunt Cycle	Zircon Channel F System II

1981–90	
Atari 5200	Atari 2600 Junior
Coleco Gemini	Atari 7800
Colecovision	NEC Turbo Grafx 16
Emerson Arcadia 2001	Nintendo Entertainment System
Mattel Intellivision II	Sega Genesis
Vectrex	Sega Master System
Nintendo Game and Watch	Nintendo Gameboy
Epoch Pocket Computer	Atari Lynx
	NEC Turbo Express
	Sega Game Gear

Table 1.3 Video Game Consoles by Decade

Table 1.3 *continued*

1991–2000	
Atari Jaguar	Sega Saturn
NEC TurboDuo	SNK NEO-GEO CD
Nintendo Entertainment System 2	Sony PlayStation Sega Nomad
Panasonic 3DO Interactive	Tiger Electronics Game.com
Sega CD for Genesis	Neo-geo Pocket
Sega Master System II	Nintendo Gameboy Color
SNK NEO-GEO	Bandai WonderSwan
Super NintendoAtari Jaguar CD	Nintendo 64
Panasonic 3DO FZ-10	Sega Dreamcast
Sega CD for Genesis 2	Sega Genesis 3
Sega CDX	Sony PlayStation 2
Sega Genesis 2	Sony PlayStation PS1
Sega Genesis 32x	Super Nintendo 2

2001–10	
Microsoft Xbox	Nintendo Gameboy Advance
Microsoft Xbox 360	Nokia N-Gage
Nintendo Gamecube	Nintendo DS
Nintendo Wii	Sony PSP
Sony PlayStation 2	Gizmondo
Sony PlayStation 3	Gamepark GP32
XaviXPORT	PDAs/Cellphones

Sources: Campbell-Kelly (2003); DeMaria (2003); Kent (2001); Thegameconsole.com (2010).

Microchips have been developed since 1971, when inventor Ted Hoff placed all the essentials needed to run a computer onto a single chip (Thurber Jr, 1995). This allowed the development of the overarching computer industry and, eventually, the flourishing of the video game business as well. Once microchips became so cost-effective that they could be used in products the average consumer could afford, which happened roughly in the early 1980s, the computer industry advanced quickly. In this period, personal computers or PCs overtook consoles and were seen as the ideal way to play video games until the resurgence of consoles in the early 1990s (Haynes, 1994). Not surprisingly, it was at this time in the mid-1980s that Atari and other firms in the console business experienced a drastic loss in sales (Cohen, 1984). It wasn't until consoles found a way to take better advantage of the continued growth in microchip power that they again became viable products (Sheff, 1993).

In the modern industry, graphic power is what is primarily achieved through microchips and, as games have approached movie quality, the power of these chips has become increasingly important (Snider, 2004). In the earliest days, however, it was processing power that made chips vital. The push to develop faster computers may have first influenced the consumer game and electronic toy market more than the personal computer industry (Nelson, 1990a). Shortages in microchips in the mid-1970s posed problems for both industries, and video games suffered much of the loss from the downturn (Business Week, 1976b; Fincher, 1978).

Because the microchip market took several years to stabilise, some analysts noted that it helped fuel the rise of software developers and publishers because they were able to evolve along different timeframes than those making consoles (Business Week, 1976c). Some microchip manufacturers were unable to fulfil even half of their orders in 1978, making development of new hardware extremely difficult. However, this meant that popular games sold out, fuelling demand at a time when the industry needed recognition and profitability (Business Week, 1979b). In contrast to the rapid development seen in game consoles, personal computers did not take off until close to the end of this period, hitting the market in roughly 1978, allowing video games another way to move from the arcade into the home.

This period of evolution for both video games and home computers pushed the development of not only microchips but also semiconductors, an industry which more than quadrupled between 1969 and 1979 (Wiegner, 1979). For instance, microprocessor producer Intel's profits increased more than tenfold in the five-year period from 1972 to 1977 (Forbes, 1977). This is particularly significant because home video game systems took several years to develop before they faced serious competition (Business Week, 1979a). By 1990, the semiconductor and microchip industries earned almost $100 billion each, due in no small part to the help of consumer electronics (Wiegner, 1979). Electronic games and early video game consoles using these technologies all became major sellers, particularly during the holidays. Companies like Milton Bradley saw electronic games come from nowhere to become a major part of their revenues (McQuade, 1979). Analysts of technology have broken these technologies into generations, identified by particular features. However, recent analyses have tended to divide generations of video game technology into the time period of their release, putting current technologies like Microsoft's Xbox 360 and the Nintendo Wii into the seventh generation of technology. However, viewing the industry as Kent (2001) and Campbell-Kelly (2003) do, in terms of style of console, affords some very useful insights. Figure 1.2 provides a comparison of the rise of video game technologies by generation, building on the initial two-generation categorisation used by Kent and Campbell-Kelly. It clearly shows a decline in consoles that coincides with the rise of the personal computer and with the crisis in software development discussed in relation to Atari and the overall industry.

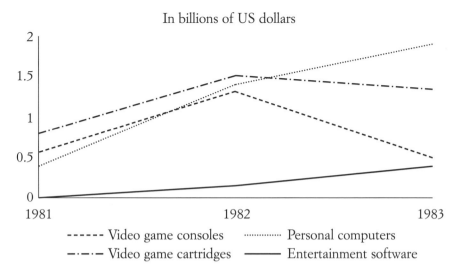

Figure 1.2 US Retail Sales of Video Games and Home Computers
Source: Campbell-Kelly (2003).

Once microchip technology took off, however, it became possible to make smaller, commercially viable platforms for home video games. In the late 1970s and early 1980s, console platforms and PCs took the lead in the market (Thurber Jr, 1995). It is significant that, as PCs became prominent in the 1980s, the video game industry began to lose its reputation as a fad. As PCs became increasingly common in homes, games became one popular use for them. One of the early selling features of low-budget PCs, like the Commodore 64, was that, in addition to their office and business functions, they could be used to play a wide variety of games (Reed and Spencer, 1986). This convergence between games and computers would grow more pronounced as both industries evolved, becoming one of the defining features of the later epochs. Though they were not popular enough on their own to push the industry to the profitability it would see in the early 1990s, it is the rise of the PC game that kept the industry remotely viable during the protracted sales slump in the 1980s (Lucien, 2002). It was not until the early 1990s that console platforms again became viable in the marketplace.

Two factors contributed to the re-emergence of the console. The first was the increased availability of microprocessors and their continued miniaturisation. This enabled the creation of both smaller, faster home consoles and an entirely new form – the handheld platform. In fact, by the early 1990s most consoles had reached a level of complexity in which their processing power was similar to the power of many mid-level PCs (Rogers, 1990). Handheld game platforms such as the Atari Lynx or the Nintendo Gameboy became extremely popular, and more developed handhelds continue to thrive today and foreshadow the development of mobile and smart phones as game platforms, though consoles and PCs dominate the market.

The second factor, however, is more significant. Beginning in the 1990s, video game hardware manufacturers finally found themselves in a market in which they were able to more successfully complete regular upgrades and in which consumers accepted these. This allowed platform makers to periodically upgrade products, for example, moving from the Sony PlayStation to the PlayStation 2, in order to benefit from new advances and to ensure a continued stream of sales. Reliance on planned obsolescence through changes in the hardware sector has given the industry two advantages – a continually changing framework for design and innovation and a regular source of restimulating dying markets (Reed and Spencer, 1986). The inability to exploit this earlier not only accounts for many of the early problems, but also suggests one of the difficulties analysts of the industry may have had in predicting its success or failure.

Software

> It takes a lot of cash to build a $20 million inventory for a three-month selling season.
>
> Nolan Bushnell, founder of Atari
> ('Atari Sells Itself to Survive Success', 1976, p. 120)

The second major commodity of the industry, software, has experienced some significant changes as well. But it should be noted that a number of key features have remained virtually constant since the beginning. The most important has been the categories of the content or genres. The early industry very quickly developed categories similar to those of today. While technical quality was obviously poorer, the range available belies the portrayal of video games as simply violent entertainment. One early breakdown of genres from 1975 included five categories, demonstrating a marked similarity to those in use today:

• combat games
• shoot/targeting games
• driving games
• artistic/maze games
• sports games. (Fincher, 1978)

As will be seen in Chapter 4, the sports genre is one that has played a particularly significant role in the success of a number of major industry players.

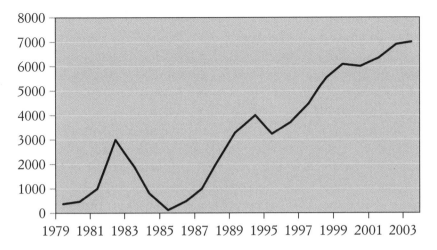

Figure 1.3 US Video Game Software Sales in Millions of Dollars, 1979–2003
Source: ESA (2002).

Genres have also become important in how audiences are cultivated, with the industry relying on key genres to maintain core audiences and sales for particular demographics. It is also important to realise that the chief historical trend in software is seen in terms of its perceived realism, or how lifelike the game feels. This is, of course, intrinsically linked to the growth in processing capabilities of the hardware.

Like the hardware sector, the software sector has experienced profound gluts in sales. As noted in the Atari example, problems in software development have been linked to some of the biggest downturns. Figure 1.3 traces the historic sales for video games in the US. The loss of sales it shows is mirrored in the contraction of console production seen earlier in the discussion of hardware.

The most important trend seen in the software industry, however, is that towards consolidation and control in production. This resulted in the emergence of a sector of publishers in the industry whose function was to guarantee game quality to the various platforms. This then extended into the marketing of game products. Early software developers tended to be limited not by their ability to create the programmes but to package and market them ('Atari Sells Itself to Survive Success', 1976). Similarly, there were difficulties in finding suitable retail outlets for their products (Cohen, 1984). The combination of these two factors resulted in a system in which distributors were able to consolidate control over the industry. A number of sources suggest this model was drawn from the Hollywood system of production in which the locus of power is distribution (Brandt, 1987; People, 1985; Wasko, 2003).

From the mid-1970s on, software development in the video game industry has worked under a two-tier structure of publishers and developers. In the late 1970s and early 1980s, the rise of independent developers did little to further the interests of the publishers, who had little control over the range of products being developed. This allowed a range of small developers whose games varied greatly in quality to force their way into the marketplace. Unfortunately, this initially had the result of flooding the market with goods, like Atari's *E.T.*, that were often knock-offs of other games at best and, at worst, were not compatible (Cohen, 1984). It was this glut of low-quality goods that put the industry into a tailspin in the 1980s, threw Atari into rough waters and which needed to be dealt with institutionally in order for the industry to stabilise.

Software plays a key role in the industry. As noted already, the planned obsolescence of hardware allows for continued development of new software. This process is costly and is, in most cases, more than any one company can hope to manage. No matter how advanced a platform's development, a failure on the software side can be fatal to the whole. As will be seen in Chapter 2, the diversity of the software sector is one key factor in avoiding such problems, while a consolidation of the hardware sector with software publishing has also helped to insulate the industry. This combination helps to allow new software to both take advantage of technical advances as well as to help justify the continued production of new, more sophisticated hardware for consumption.

Audiences

> Companies looked at the market and recognized that kids aren't the only ones
> playing with toys.
>> Milton Schulman, former editor of *Toy and Hobby World* (Rice, 1979, p. 97)

Perhaps the most important trend, however, pertains to the audience for the industry's products. While it has become more common to see people discuss older audiences for video games today, there has always been a high percentage of adult buyers. Just as video games began in arcades but ultimately found a sort of legitimacy by moving to personal computers and consoles, so, too, has their audience expanded.

In the early days, the players were experienced computer programmers. Because games required mainframes available only to government and university researchers, the players were, of necessity, educated adults. As games moved into arcades, the audience changed. In a classic example of 'early adopters', children and technologically inclined adults adopted video games as a form of entertainment (Fidler, 1997). This shift resulted in a decline in the importance of age and education, but introduced income for discretionary spending as one of the key factors in who played games. The advent of home consoles continued this trend. Early consoles, particularly in their first phases of development, could cost several hundred dollars.

It's with the shift to arcades that the first seeds of the idea of video games as a product for children came to prominence. The view that arcades were primarily intended for children and adolescents was already well established in the popular press. In contrast, the first PC games were dependent upon highly skilled users, and early systems were considerably more expensive, often relying on one's ability to assemble the computer oneself, making the initial audiences limited (Condry, 1984).

That these trends somehow allowed for video games to be treated primarily as children's products is interesting, but it did allow for dismissal of the industry in the popular press prior to the early 1990s. The popular view of the industry through the 1980s was that the audience for games was relatively limited, and so undeserving of serious consideration.

This had another interesting effect, however. As long as investors saw video games as having both a high cost and a market restricted primarily to children, they remained skittish (Tapscott, 1997). At least some of the explanation for the long road back to success lies in this perception of a limited audience. Not just investors but some companies began to believe it as well. If most games were targeted towards the youth audience, not only was there the danger of exhausting the sizeable (for that age) but limited discretionary income. In order to guarantee continued sales, a wider audience had to be found. It was up to the industry to convince the world that it existed. Technology would serve as the primary answer. When technological advances allowed the industry to shift from one console generation to the next, advances in graphics allowed for more involved stories, and these were touted as a means to draw in the elusive adult audience (Kent, 2001).

The reality is somewhat different. As early as 1979, the industry was targeting products not just at children but at adults or entire families (Business Week, 1979b). It was not unusual to find that even products directly aimed at the youth market became very popular with adults (Business Week, 1979a). Milton Bradley reported that more than half of its electronic games were purchased by people between fifteen and thirty-five years old (Business Week, 1979b). This may explain why when Milton Bradley launched its Comp IV video game system, it chose to market it in Manhattan bars while its popular game *Simon* was introduced at an expensive bash at Studio 54 (Rice, 1979).

The trend became so pronounced that interior designers in several large cities were approached to design entire rooms dedicated to electronic games. Of course while Milton Bradley produced primarily electronic games rather than video games, other companies focused primarily on the latter, with Atari reporting similar trends (Business Week, 1979a). In fact, one estimate suggested that as many as half of all video game purchases in 1979 were by adults for adults. As Milton Schulman, then editor of *Toy and Hobby World*, put it in 1979, 'This is the start of a whole new ball game. Companies looked at the market and recognized that kids aren't the only ones playing with toys' (Rice, 1979). By 1990 adults purchased a considerable portion of Nintendo's products, always known for being heavily youth-oriented (Moffat, 1990).

Another interesting point is that demand was so high for early video games that prices were able to stay fairly high. In 1976, when there were only two major console producers, Atari and Magnavox, and the industry was just emerging, console prices typically ranged between $60 and $100 ('Demand Overwhelms Video Game Makers', 1976). However, the market was still extremely limited, with products being sold almost exclusively in the US and Japan. It wasn't until around 1990 that many video game companies began to sell actively in Europe (Moffat, 1990).

Even so, production of games themselves – the software side of the industry – was in full swing. Though the market was still growing, more than 100 titles were produced in the early boom years (Ferris, 1977). Exact numbers are difficult to obtain because the services which today track the industry, such as the Electronic Software Association (ESA) did not exist at the time and have little data before 1998 (Hewwit, 2005). The industry today continues, however, to try to make in-roads with new audiences and relies heavily on the adult audience for its profitability, as discussed in Chapter 2.

Two companies provide excellent examples of how the industry began to emerge from the struggles experienced by Atari. The first is Nintendo, whose advent heralded the modern industry, with tight control of production and licensing and by pushing convergence and development of new platforms. In contrast, Sega emerged as both a major console manufacturer and software producer in the 1980s, only to falter and pull out of hardware by the end of the 1990s.

CASE STUDY: NINTENDO

By 1990, the video game market had stabilised. However, the reasons for its resurgence were not because of the typical myths about the industry. Manufacturers were reporting steady growth not just in the number of adult players but in female players as well (Rogers, 1990). In fact, nearly 30 per cent of players were twenty-five years old or older. More than 400 software titles were available, twice the number of two years earlier (Wheelwright, 1990). The design

Year formed	1933
Headquarters	Kyoto, Japan
2011 Sales (US$m)	12,240.1
2011 Employees	4,712
Industry sector	Hardware, software publisher
Hardware and year introduced	
Consoles	Color TV Game – 1977
	Nintendo Entertainment System (NES)/Famicom – 1983
	Super Nintendo Entertainment System (SNES) – 1991
	Virtual Boy – 1995
	Nintendo 64 (N64) – 1996
	Gamecube (2001)
	Wii – 2006
Handhelds	Gameboy – 1989
	Gameboy Advance – 2001
	Nintendo DS – 2004
Key software franchises	*Mario Bros.*
	Donkey Kong
	Pokémon
	Tetris
	Nintendogs
	The Legend of Zelda
	Starfox
	Yoshi
	Wii Sports
	Wii Play
	Wii Fit
	Brain Age/Brain Training

Table 1.4 Corporate Profile of Nintendo, Inc., 2012
Sources: Hoover's (2012); Mergent (2008).

of games themselves had stabilised, and the industry's structure had begun to solidify, with individual workers starting to specialise in particular aspects of production, while the relationships between companies involved in hardware production, software production and distribution took on the form they have today (Netsel, 1990).

Founded in 1889, Japanese company Nintendo has worked its way from an arcade giant to one of the leaders in video game platform production. Its Gamecube console was battling with Microsoft's Xbox for second place globally, behind Sony's PlayStation systems. But its Gameboy systems, including

2004's Gameboy DS, lead the handheld sector. In addition, Nintendo publishes video game software, and has focused on the children's market (Hoover's, 2005; Nintendo, 2005). The company's first foray into global console gaming was in 1985, with its NES system, which represented a key event in revitalising the industry after a serious sales slump. Nintendo released its first handheld, the four-colour Gameboy in 1989, and dominated the handheld sector until Sony released its PSP in 2004 (Nintendo, 2008b). Nintendo appeared to have lost ground during the sixth generation of consoles, with its Gamecube falling into a distant third place behind Sony's PlayStation 2 and Microsoft's Xbox, but with the seventh-generation console, the Wii, Nintendo has become one of the power-houses again (Hoover's, 2008), maintaining facilities in Japan, Spain, the US, Australia, the Netherlands, Belgium, the UK, Italy and Hong Kong (Nintendo, 2008a).

By 1990, the US market for video games had reached approximately $2 billion, a leap of almost 30 per cent from previous years (Brandt, 1990); globally, the industry was worth almost $10 billion (Neff and Shao, 1990). The US and Japan were the primary markets while Europe was still treated as a secondary market, receiving products up to three years after the primaries (Moffat, 1990). Household penetration in the US had reached nearly 20 per cent while Japan had surpassed 30 per cent (Shao, 1989). It was Nintendo that would take charge of expanding the video game market worldwide, establishing a subsidiary in Frankfurt (Moffat, 1990).

Nintendo was noted for its forward-thinking marketing and tight control of its property. Starting as a manufacturer of playing cards, Nintendo has grown considerably from its humble origins (Nintendo, 2005b; Sheff, 1993). By the mid-1950s, the company had expanded into arcade games and from there it was a short leap into video games. But much of its focus was on the Asian market, and it was not until the 1970s and 80s that it began to expand into other markets (Provenzo Jr, 1991; Sheff, 1993).

Nintendo's major impact on the US video game market was much more recent. Its first console, the Super NES and games for it hit the shelves in 1986. By 1990, Nintendo consoles and software made up 90 per cent of US sales. When game sales are counted as part of toy sales, Nintendo made up almost 21 per cent of total toy sales for the same year (Peterson and Shao, 1990). The company had US sales in 1990 of between $2.5 and $2.7 million, roughly ten times its sales a decade earlier (Time, 1990; Moffat, 1990). Estimates vary, in part, because of the exchange rate of yen to dollars. In the short period between 1986 and 1990, the company sold over 40 million consoles worldwide (Nelson, 1990a). Globally, Nintendo accounted for more than 80 per cent of worldwide game sales (Moffat, 1990).

Sitting at the top of the market would seem to be a blessing, but for Nintendo it spelled trouble. Having proved that there was a path to success in the video game industry, competitors began to watch the company's tactics closely. More than one analyst suggested that, with Nintendo having reached such impressive household penetration, it was possible that demand may have reached the saturation point (Moffat, 1990, Business Week, 1990; Peterson and Shao, 1990). Since then, even though the company's control over the market has slipped, it has maintained a strong presence, particularly in the area of portable consoles. The original Gameboy, introduced in 1989, ultimately sold more than 118 million units worldwide by 2010 and has spawned a number of successors and competitors (Biersdorfer, 2004).

What made Nintendo successful was more than just its technology or its grasp of the market. Certainly the products were popular; in 1990, for example, the company agreed to a petition from the Japanese government to release new games only on Sundays as a means of discouraging truancy (Nelson, 1990b). More significant was how much effort the company put into repositioning its products from fringe toys or fads to products which were seen as part of normal, everyday life. Its advertising efforts, in particular, were central to this. Though certainly not the first in the field to recognise the importance of advertising, Nintendo took advantage of advertising power earlier and with more zeal than any of its competitors. It is estimated that the company spent upwards of $60 million in advertising and promotion in 1989 (Shao, 1989). In addition, it succeeded in finding ways to expand its market and control the entire process of game production. By 1989, the company partnered with Toys'R'Us to create 'the World of Nintendo', a shop within a shop, with all manner of products including games, t-shirts, toys and other paraphernalia (Shao, 1989). At the 1990 Consumer Electronics Show, Nintendo's booth was not only the largest, in keeping with its sales, but also featured one of the strangest arrays of products from video games to action figures to breakfast cereals (Rogers, 1990). The cereals produced by Ralston Purina featured prominent Nintendo characters and sold well into the 1990s. The company also had a deal with PepsiCo to market its brand Slice using Nintendo's *Super Mario Bros.* (1985) (Shao, 1989). As a franchise, 'Mario Bros.' is one of the greatest triumphs of the video game industry. The property was spun off from an early Nintendo game, *Donkey Kong* (1981), and has sold more than 39 million copies worldwide, spawned numerous video games and a movie. One poll of US schoolchildren showed Mario himself to be more popular than Disney's Mickey Mouse (Moffat, 1990).

The company also spent considerable effort making its consoles more than just machines for games. Recognising the potential of the rapidly advancing processing power of consoles, Nintendo began to experiment with networking its machines. By 1989, it was possible to connect a Nintendo to a modem and use it to check stock prices and financial information (Rogers, 1990).

Nintendo's control over the products played on its systems was also impressive. Research and development of in-house products accounted for roughly 10 per cent of Nintendo games; all other products were developed by licensed affiliates, which were also responsible for marketing costs even though Nintendo maintained the right to veto any game from being shipped. Firms with an idea had to develop the game according to specifications from Nintendo, get the company's approval, pay for the cost of cartridge manufacturing and agree not to supply the game to anyone else (Moffat, 1990). Moreover, the company maintained strict control over the technology for making both consoles and cartridges as a way of minimising the threat of third-party publishers encroaching on the market (Pitta, 1990). So restrictive were the licensing agreements that even Electronic Arts, which had been selling games for fifteen platforms, only licensed eleven of its more than 350 video games for Nintendo systems in 1990 (Pitta, 1990).

The company's control on production was so rigid that it drew attention from outside the industry as well. Nintendo's licensing policies became a sticking point in US and Japanese trade agreements in 1989. Congress created a subcommittee to investigate the practices, ultimately resulting in both the Department of Justice and the Federal Trade Commission beginning preliminary investigations (Moffat, 1990). It was during these investigations that Nintendo's grip began to slip. Companies began to reverse-engineer, releasing games unlicensed for Nintendo systems. Among those companies were the still floundering Atari and the ascendant Electronic Arts.

The study of both Atari and Nintendo demonstrate how long the video game industry's current structure has existed. These policies resulted in extreme concentration of ownership, tight control of intellectual property and the need to continually expand markets. Nintendo's business practices also represented a significant shift in the global nature of video game production, as the company sought to make its products more marketable globally, rather than targeting primarily Japanese audiences (Iwabuchi, 2002). There is also evidence of labour practices that, as will be seen in Chapter 5, still exist today. These practices have resulted in a strict division of labour and have limited the effectiveness of workers to influence production. Finally, they also demonstrate the significant and long-standing ties between the video game industry and other media industries.

CASE STUDY: SEGA

Unlike Nintendo, Japanese company Sega, which was formerly a hardware producer but ultimately shifted its focus to software publishing, has had a rougher

Year formed	**1951**
Headquarters	**Tokyo, Japan**
2012 Sales (US$m)	**4,806.5**
2012 Employees	**6,700**
Industry sector	**Software publisher, former hardware manufacturer**
Key hardware franchises	
Handhelds	Game Gear – 1990
	Nomad – 1995
	Mega Jet – 1994 (Japan only)
Consoles	Saturn – 1995
	Dreamcast – 1999
Key software franchises	*Sonic the Hedgehog*
	Bayonetta
	Crazy Taxi
	Virtua Fighter
	Super Monkey Ball
Studios owned	Secret Level – San Francisco, California, US
	Creative Assembly – Horsham, West Sussex, UK
	Sports Interactive – London, UK
	Prope – Tokyo, Japan

Table 1.5 Corporate Profile of Sega Sammy Holdings, Inc., 2012
Sources: Hoover's (2012); Mergent (2008).

time in the industry. Despite its struggles with hardware, Sega has consistently performed well in Asia and Europe. Table 1.5 provides a brief breakdown of Sega's current circumstances, including its major franchises and hardware offerings. Sega was founded in 1951, focusing on importing coin-operated photo booths and jukeboxes but eventually then coin-operated games into Japan (Hoover's, 2010). Its primary initial market was US military bases overseas, but the company recognised the need to expand (Sega, 2010). Within a few years, it had expanded its purview throughout Asia and into Europe (Hoover's, 2010). The company did well for itself, and merged with an American firm, Rosen Enterprises in 1965, helping its breakthrough into the US market. Its first major arcade hit was a game called *Periscope*, released in 1965 (Sega, 2010). By the mid-1960s, Sega had moved into manufacturing its own coin-operated machines, after acquiring a jukebox manufacturers (Hoover's, 2010).

Eventually, the company was profitable enough to attract outside interest, and in 1969, it was sold to Gulf & Western. It maintained its profitability, drawing primarily on arcades, though it had begun to expand into other areas, and was drawing profits of $214 million in 1982 (Sega, 2010). The company began to experiment with consoles, introducing its first console in Japan in 1983, the SG-1000, as well as video game software, but its primary interests remained in arcade production (Kent, 2001). When the arcade business began to die off in the mid-1980s, Gulf & Western sold Sega to rival arcade game company Bally Manufacturing Corporation in 1983.

In Japan, however, Rosen was able to purchase the Japanese assets of Sega for $38 million, and the company diversified its practices (Sega, 2010). It continued its successful push into arcades and hardware manufacture, becoming one of the leading concerns in both Asia and Europe and establishing arcades and manufacture in the US. By 1986, the company was doing well and established Sega of America to adapt its products to worldwide audiences, including the lucrative American marketplace.

As the video games industry began to struggle in the mid-1980s – particularly arcades – Sega began to work harder on consoles and handhelds, recognising the importance of children as a market (Kent, 2003; Sega, 2010). In 1985, it released its first console intended for worldwide audiences, the 8 bit Sega Master System (Kent, 2001; Thegameconsole.com, 2010). The console struggled to sell, losing to Nintendo's NES system (Hoover's, 2010). Much of the competition centred around the additional features of each console. (Inside the Internet, 2000). As will be seen in Chapter 3, however, the console market, though concentrated, is subject to volatility when the limits of one generation's technology are reached. Sega's next release, the Sega Genesis (in some markets, referred to as the Sega Mega Drive) captured both a high share of the US market and the hearts of many game players, who still seek out its games today. Unfortunately, it was much less popular in Japan, and ultimately, Sega moved quickly towards the development of its next system, the Sega Saturn (Kent, 2001). The Saturn, however, had the opposite problem: it was popular in Japan and Europe but sold poorly in the US, losing to Nintendo's N64 and the PlayStation, which went on to become the leading console in the world. One recent estimate indicates that, while the Saturn has sold almost 6 million units worldwide since its release, the PlayStation, the first console from Sony and competitor to the Sega Saturn, has sold more than 19 million.

Sega attempted one more foray into the console industry, releasing the Sega Dreamcast in 1998 (Olenick, 1999). It dominated sales, until both Sony and Nintendo released their next consoles – the PlayStation 2 and the Gamecube (Inside the Internet, 2000). In addition, Microsoft began development of its first console, the Xbox. The intense competition was too much for Sega, and it withdrew from console production in 2001. Sega did not completely abandon hardware, partnering with Nokia to develop a cellular phone that could play video games, presaging the rise of smart phones in today's market (Guth, 2003; Lou, 1999; Pringle, 2002). Ironically, the price drops on the discontinued Dreamcast drove up demand, a moment which hinted at the strategy used by current hardware makers of dropping prices and treating their consoles as loss-leaders (Huebner, 2002).

Sega-Sammy Holdings emerged from the 2004 merger with Sammy, the country's top producer of pachinko and pachislots – mechanical gambling games popular in Japan – to become a top game producer in Japan. The company also owns substantial properties in amusement parks in Japan (Hoover's, 2008) and maintains facilities in Japan, China, the US, the UK and Australia and formed ties to Lucas Arts and Sony (Sega, 2007).

Sega's example is one of weathering the volatility of an industry. Over its history the company has been a major player in every sector, surviving by successfully changing direction in times of crisis. When its foray into hardware flagged, the company refocused, selling off a number of its subsidiaries, including its amusement park division, in order to concentrate on the creation of software (Hoover's, 2010). By producing software for the remaining console makers, Sega has experienced considerable success as a publisher.

CONCLUSION

From the earliest days of video games, the potential for the business to become a vital part of the cultural industries has existed. Early companies consciously modelled themselves after successful firms in other industries and then sought out connections with them. By taking this path, video games entered cultural discussions not as works of art but as commodities. By the mid-1990s, a formal, recognisable industry had emerged, distinct from other cultural industries, though inextricably related to them. In a few short years, the industry had moved from being an ancillary market to other, well-established industries like television and film to representing a significant rival. By recognising that video games have been commodities from the beginning, it becomes possible to draw valuable conclusions about their production as well as their function and impact on society.

The first of these conclusions is that video games, both in function and design, have always been more than toys. Their production owes much to the toy industry, particularly its production schedule. It also draws heavily from the computer industry and, as will be seen in later chapters, the motion picture industry. From computers, video games adopted the use of rapidly advancing technology and planned obsolescence in order to keep its products viable. Moreover, the distinct production of video games as software is a recent phenomenon. A number of important companies in both computer software and video game software have only in the last fifteen years begun to fully differentiate, developing segments across all sectors of the industry while extending into other media. This accounts for the similarities in production and attitudes to labour seen in Chapter 5. Similarly, the video game industry has modelled the control of its products, its relationships with retailers and advertisers, as well as how to best distribute them, on the Hollywood system. As the industry has developed, it has recognised the need for both synergy and advertising. The example of Nintendo is particularly instructive on these points. The current structure of publisher–developer relationships discussed in greater length in Chapter 2 emerged early on in the history of the business. Like Hollywood, the industry has sought to retain tight control of its products while attempting to find new ways to synergise them more effectively. These attempts are examined more fully in the next chapter.

The second conclusion suggested by this history is that the industry has long acknowledged how untenable its existence would be if supported only by the discretionary spending of children. From its earliest days, the industry has courted adult audiences. This has helped it to survive a number of market crises as well as reinforcing its distinction from the toy manufacturing industry. Where the industry has been traditionally weak has been in courting the global market, something that it has only begun to address successfully in the last decade.

Finally, because video games emerged from the development of computers as a means to train and demonstrate the capabilities of available hardware, they emerge as devices of communication, useful in support of, and dangerous because of competition to, other media industries for audience time. Further, because video games present examples of some of the strongest attempts at controlling content even as they are increasingly combined with technologies that endanger this control (such as networked consoles, development tools usable by players, etc.), their role as intellectual property – and as forms of communication – needs to be better understood and so is taken up in later chapters. That video games' communication capabilities are only now being truly explored is a loss, but it is due in part – as with so much in the industry – to the limits of the available technologies.

2

Market Structure, Audiences and Software Production

As the modern industry cemented its structure in the mid-1990s, software was clearly the most important sector, driving profits as hardware manufacturers had resorted to treating their consoles as loss-leaders. In addition, the industry had begun to formalise ties with outside industries like film and television to bolster their productions, tying in to established franchises and their production schedules, to help minimise the risk involved in developing new software (Nichols, 2008). By moving away from production that was dictated by these industries, video game software production helped establish video games as a distinct cultural industry. This chapter focuses on establishing the broad structure of the modern industry and on the cultivation of audiences and the development of software as central to its success. The matter of hardware production is taken up in Chapter 3.

In the broadest terms, cultural industries must deal with both the production and distribution of content. Each develops its own unique formulation for production, based on historical precedent. For example, the movie business is organised into a system of production, distribution, exhibition and retail while recorded music has adopted a production, promotion and distribution format (Frith, 1997; Wasko, 2003). Many media industries follow similar structures with their own unique adaptations. The video game business developed a structure similar to the preceding examples but adapting it to the logics of a high-tech, information industry. What this means is that, unlike film or recorded music, the video game industry has incorporated as one of its logics of production the need to keep up with technological advances. In fact, it has incorporated such advances into its business model as a means of keeping profits up. In contrast, the film industry has been shown to be slow to adopt new technologies and, as with digital cinema, has not found a way to make the transition easily affordable (Wasko, 1994). Similarly, the music business has struggled to adapt to advances in technology, resorting to increased legal action to maintain its foothold (Reuters, 2000a, 2000b; RIAA, 2000).

This chapter begins the examination of the modern video game industry by detailing its overall market structure. It then analyses its audience and attempts to dispel some of the myths about who plays and who doesn't. It discusses the software sector and how the games we play are produced and distributed. It provides case studies on two of the major game publishers. It concludes by examining software genre sales data and the difficulties and ambiguities of regulating software.

MARKET STRUCTURE

As noted in Chapter 1, the industry's reliance on incremental advances in technology is a key feature in its development and profitability plans. Advances in graphics capability, storage capacity and convergence of technologies have been harnessed to create periodic reissues of hardware. Typically this has fallen into a cycle of from two to five years, with the major players in the hardware sector releasing new equipment with new specifications for software developers to design to, including new capabilities in depiction and gameplay. It is at these moments of transition that the hardware sector becomes most volatile, allowing new companies a chance to come to prominence, as seen in the example of Sega in the previous chapter. This, in turn, renders the older models relatively obsolete though, as will be seen with the example of the Sony PlayStation 2 in Chapter 3, it is not always enough to force older versions out of the market entirely. Changes in software are largely how audiences recognise the advances in technology.

Further, the transnational nature of the video game industry cannot be stressed strongly enough. Currently there are three primary sales regions – North America, Europe and Japan, though the East Asian market is growing in importance (Dyer-Witheford, 2002). The major centres for software production are also primarily in these areas, as demonstrated in Figure 2.1. The figure, drawn from a self-reporting website, shows a clear concentration of game publishers and developers in the major markets for video game consumption. In Japan, almost 80 per cent of households own and play video games (Aoyama and Izushi, 2004), making the country the second largest market behind the US, and it has served as a first market for many companies' game development and testing (Guth, 2001). Moreover, it is home to two of the biggest games companies, Nintendo and Sony, both producers of hardware and publishers of software. The Japanese market is also notable because it has proved a remarkable test of global game success (Nelson, 1990a).

In contrast, the European market has tended to be treated as the final resort and has been – as was noted in Chapter 1 – often the last market for new hardware to be deployed (Ip and Jacobs, 2004; Nielsen, 2005). But Europe has become increasingly important, and the growth in audiences there has not only drawn more production to the region but has also resulted in attempts to

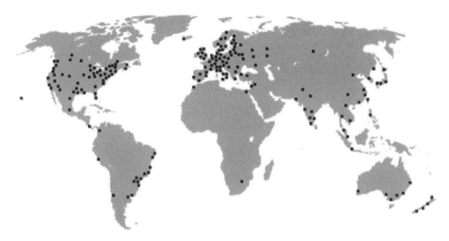

Figure 2.1 Map of Game Developers Worldwide, 2010
Source: Gamedevmap (2010).

establish government subsidies for game development (Gamasutra, 2005b; Leisure and Travel Business, 2010; Nielsen, 2008b). Korea has also proved an important market, with a large base of involved players, particularly for both PC and mobile games. One estimate suggests that as many as 1,200 different companies are involved just in online gaming in South Korea (Ihlwan, 2007). It is estimated that almost 80 per cent of households have at least one computer, and this has led to a strong industry worth $1.7 billion in video game sales in 2006. This dwarfs estimates of China's game market in the same period, which earned approximately $760 billion (Caldwell, 2006). Video gaming in Korea has become so popular that professional players are accorded superstar status (Schiesel, 2006).

This suggests two key areas of the video game industry structure: hardware and software manufacture. Hardware manufacturers can be broken down into particular types such as consoles, handheld machines, arcade games, peripherals (like game controllers), etc. Moreover, the development of software, the most lucrative area of the business has a distinct structure, encouraging rapid development of products by small firms, which are then distributed or published by larger firms, which also absorb most of the profits. Figure 2.2 illustrates the relationships of the modern video game industry. The emergence of publishers, who guarantee compatibility and quality to the hardware manufacturers, is one of the key developments. Publishers developed in direct response to the crisis that production companies like Atari faced in the early 1980s. They enforced this quality via strict controls, such as those Nintendo was noted for in the 1990s. As a development, publishers function similarly to distributors in the film industry, which is not surprising considering at least one of the major game publishers, Electronic

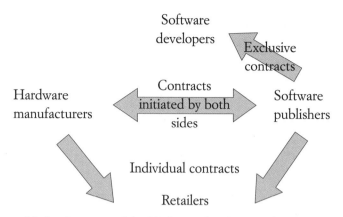

Figure 2.2 Market Structure of the Modern Video Game Industry

Arts (EA), modelled itself on a Hollywood company. Electronic Arts is profiled later in this chapter. In addition, the industry has developed extensive networks for promotion (and cross-promotion) and retail agreements, which most commonly (though not always) are controlled by game publishers. It is also important to stress, as noted in the introduction, that this structure describes the mainstream industry, and that alternative forms of production, such as independent or art games, exist both in response and alternative to this structure.

As noted previously, video games are sometimes considered a special category of toys. Within the toy sector, however, they are the fastest growing segment while the rest of the industry has experienced stagnation in recent years (Kang, 2005). Already, the $20.3 billion US toy business is struggling, with video games making up more than half of the industry's profits (Elkin, 2003). The reach of video games is only expected to grow with recent predictions suggesting that they will earn more than $50 billion globally per year (Grover *et al.*, 2005; Marr, 2005).

Globally, the impact of video games is more impressive. Global revenues for the top game companies in 2004 grew 71 per cent from the previous year to $41.9 billion (Bond, 2008; Gamasutra, 2005d). By 2009, the industry beat these predictions despite slow sales growth owing to global recession. In 2009, total sales globally had reached more than $77 billion (Graft, 2010a). Estimates suggest that this might increase to more than US $82 billion by 2017 (Gaudiosi, 2012). Those patterns were established early, with the leading firms stabilising between 2003 and 2007, Table 2.1 provides a breakdown of the top four companies and their revenues from 2003 to 2007. The four shown consistently make up a majority of the industry's profits, and each is a major publisher. Of the four, only Electronic Arts (EA) is not also a hardware manufacturer. It's interesting, as well, to note that all four are headquartered in either the US or Japan.

	EA	Microsoft	Nintendo	Sony	Total
2007	3,091	6,083	9,045*	9,117*	27,336
2006	2,951	4,256	4,766*	7,848*	19,821
2005	3,129	3,140	4,819*	6,566*	17,654
2004	2,957.1	2,876	4,689	7,502	18,024.1
2003	2,482.2	2748	4,203.3	7,958.3	17,391.8
% change	18.9	4.7	11.6	(5.7)	3.6
% of industry, 2004	12.0	11.7	19.1	30.1	73.5
% of industry, 2007	7.4	14.6	21.5	21.8	65.3

Table 2.1 Revenues of the Top Four Companies in the Video Game Industry,
     2003–7

Notes: * Based on 2008 yen to dollar conversion. All figures are in millions of US dollars
and represent only revenues reported from video game-related business segments.

Sources: EA (2005, 2006, 2007, 2008); Gamasutra (2005d); Microsoft (2005, 2007);
Nintendo (2005, 2008); Sony (2005, 2006, 2008).

In the US, video game sales have stayed steady during the same period. Although there was a drop from just over $10 billion in 2003 to $9.9 billion in 2004, sales grew to more than $17.4 billion by 2007. Concentration, however, has not suffered. Domestically, the top five companies in the industry accounted for well over 50 per cent of US software sales (Grover, 1998). From 2003 to 2007, software sales grew from $5.84 billion to just over $10 billion (Peckham, 2008; Wingfield, 2005). If hardware were factored in, the number would closely mirror the global concentration figure (Grover *et al.*, 2005). The profit margin on software development has also remained healthy; in many cases, it approaches a 25 per cent margin or almost three times that of most motion pictures (Holson, 2005a).

What is perhaps most impressive about the video game industry is that it is still largely tied to a single quarter of sales for the majority of its profits. In 2003, for example, the roughly $7 billion in video game goods was sold during the Christmas holiday season alone (Richtel, 2004). In most years, the majority of these profits comes from the sale of software, particularly because many console makers are frequently required to drop their prices to help bolster sales (New York Times, 2002). In 2003, for example, it is estimated that 70 per cent of total video game revenues came from the sale of software (Snapshot Series, 2003, 2004). Perhaps not surprisingly, many of these came from adult audiences rather than from children (Richtel, 2004). Such a pattern suggests that understanding the nature of the modern video game audience is crucial to understanding the industry.

By 2012, this concentration was less noticeable, though those four companies were still out in front. Globally, video game revenues reached an estimated US

$67 billion (Gaudiosi, 2012). During that period, Electronic Arts earned revenues of US $4.14 billion, with US $1.36 billion coming in the fourth quarter of that year (Mallory, 2012). Nintendo, which had released its eighth-generation console – the Wii U – estimated that it had earned 14 billion yen – approximately US $1.5 billion – in 2012, even as sales of the new console failed to meet projections (Tabuchi, 2013). Despite being the first company to release an eighth-generation console, the slipping sales suggest a shift in console dominance as hardware competitors Microsoft and Sony were both preparing to release their own new consoles. Microsoft saw similar patterns, earning US $3.78 billion in 2012 (Sarkar, 2013). Sony, which had struggled financially following declining sales and production difficulties stemming from natural disasters near its plants in Japan, posted revenues of US $2.86 billion in 2012 (Handrahan, 2013). Based on those numbers, the big four earned approximately 18 per cent of total industry revenue in 2012. A number of factors account for the shift, including the rise of online and mobile gaming, discussed later in the chapter. Estimates suggest that, by 2015, those areas will earn as much as US $48 billion or 57 per cent of total revenue (Rose, 2013). This has led to firms like Apple and Google entering the games market, even as they promise to reshape the global makeup of game sales. East Asian countries including South Korea and China are likely to become major outlets for game sales in coming years. Similarly, shifts towards digital distribution have allowed other companies like US-based Valve to make in-roads, though how viable digital distribution will be globally remains to be seen.

AUDIENCES

As a cultural industry, one of the primary logics of production that video games must address is how to view and cultivate its consumers. As Chapter 1 discussed, the games audience has long consisted of more than just children, but simply attracting some adult interest is not sufficient for any industry, and, as such, video games are constantly seeking to expand their audience.

In fact, the audience for video games is quite impressive in its diversity. Conservative estimates suggest that as many as 50 per cent of Americans over the age of six play them, and approximately 17 per cent of computer-owning households include someone who plays one form of online game or another (Bulkeley, 2003; ESA, 2005b). These numbers compare favourably to those in Europe. According to a study conducted for the Interactive Software Federation of Europe (IFSE), 37 per cent of people in the UK between the ages of sixteen and forty-nine play games, while in Spain and Finland, 28 per cent of the population are active gamers. Moreover, game use has risen from previous years, and incorporated a variety of platforms (Nielsen, 2005, 2008). In Japan, roughly 37 per cent of the population played video games actively (Niizumi, 2004). Studies in South

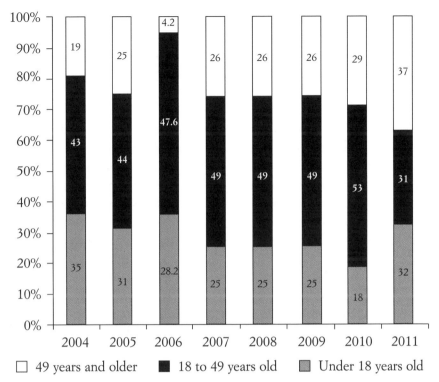

Figure 2.3 Comparison of US Video Game Player Age, 2004–11
Note: Data incomplete for years 2004 and 2006.
Sources: ESA (2004, 2005a, 2006, 2007, 2008, 2009, 2010, 2011, 2012).

Korea show that as much as a third of the country has played a single game, Nexon's *KartRider* (Ihlwan, 2007). In Australia, estimates suggest that 79 per cent of households play them (IEAA, 2007).

Not surprisingly, video games are also tremendously popular with college-age individuals. In the US more than 65 per cent of students indicate that they are regular gamers (Carlson, 2003b). However, games are also quite popular with an older crowd. In the US, the average player age had risen from twenty-nine years old in 2004 to thirty-five by 2010. Estimates also suggest that at least 17 per cent of all US players are over the age of fifty (Emeling, 2004; ESA, 2005a, 2010). Figure 2.3 provides a breakdown by age of US video game players, showing the change from 2004 to 2011.

Internationally, similar age breakdowns have also been found in Europe. The average player in Spain is twenty-six years old; in Finland, thirty and in the UK, the average player is thirty-three (Nielsen, 2008b). A 2003 study in Great Britain showed that only 21 per cent of the country's gamers were children or teens. Almost 80 per cent were aged twenty or over, and almost 16 per cent of players

were over thirty-five (Emeling, 2004). A study by the BBC confirmed this, showing that almost 60 per cent of people between the ages of six and sixty-five in the
UK are regular game players (Pratchett, 2005). A similar pattern can be seen in
Australia, where the average age of players was twenty-eight in 2007, up from
twenty-five in 2005 (IEAA, 2005, 2007). Globally, the average age for a video game
player in 2005 was thirty-five years old (Wingfield and Marr, 2005). Interestingly,
older gamers tend to report an attraction to computer games as a means of building social networks between generations, an area ripe for study.

Having ensured popularity with both young and old, the industry has also
taken steps to bring female gamers into the fold as well. One strategy has been
to base games on branded media properties that have proven popular with
females. One example is the series of games based on Disney's *Lizzy McGuire*
(2001–4) and Fox's *American Idol* (2002–) (Swett, 2003). Such moves have had
some success. Girls aged six to seventeen make up roughly 12 per cent of the
total US video games market, while women over eighteen make up a full 26
per cent (Loftus, 2003). Figure 2.4 breaks down US video game players by gender from 2004 to 2011, while Figure 2.5 provides evidence that female gamers
are an active part of most global markets. Since that time, studies show that
already as many as 39 per cent of all gamers are women (Grover *et al.*, 2005).

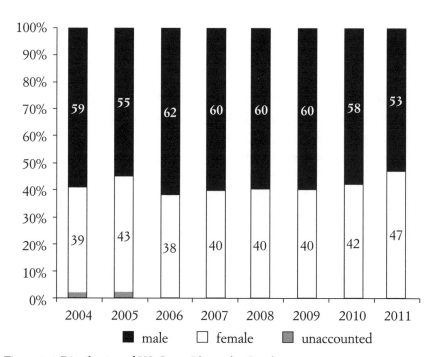

Figure 2.4 Distribution of US Game Players by Gender, 2004–11
Sources: ESA (2005a, 2006, 2007, 2008, 2009, 2010); (ESA, 2011, 2012).

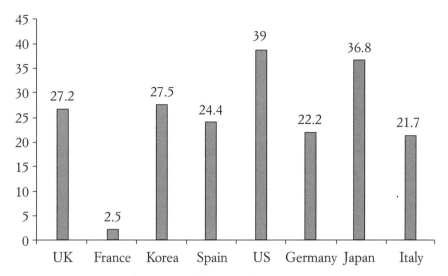

Figure 2.5 Percentage of Active Female Gamers by Country, 2003
Source: ELSPA (2004).

This number is expected to continue to grow as casual gamers, those who tend to play basic games requiring little time commitment or game knowledge often on platforms other than consoles such as on cell phones and social networking sites, become more common. Estimates in the US show the number has continued to grow, with 2010 estimates indicating at least 40 per cent of all gamers are women (ESA, 2010). In Europe, such breakdowns are harder to find, but women who describe themselves as active gamers are on the rise. In Spain and Finland, nearly 18 per cent of all active gamers are female, while in the UK the number has soared to 30 per cent (Nielsen, 2008b). In the Spanish example, gaming is becoming an important aspect of media consumption by females, with more than 54 per cent of surveyed women playing (Perez Martin *et al.*, 2006). Australia also shows a high rate of female gamers, with an estimated 41 per cent of players being female (IEAA, 2007).

The chief differences between male and female gamers seems to be which sort of games they play and how, in addition to how much they play. This is seen in questions about whether gamers consider themselves active game players or not. Studies show that women tend to play more games on PCs and the Internet than men, while they make up almost an equal share of console players (Carlson, 2003b). Women also seem to prefer games, particularly online games, which are less competitive and more socially oriented (Roe and Muijs, 1998; Subrahmanyam *et al.*, 1999). The industry has been only too glad to grab onto the trend. One online game site, Real Networks, reported that approximately 70 per cent of its users were female, paying $6.95 a month for access to the game site. Competitor Lycos indicated that more than 70 per cent of its subscribers were women (Swett, 2003). In 2006, a Nielsen study indicated that as much as 64 per cent of all online gamers in the US were women, suggesting a growing market segment for game producers to take advantage of (Marketingvox.com, 2006).

Developers have begun to target older players in similar ways. Beyond trying to find television and film synergies, they have begun to experiment with ties to music labels. Some features of these deals have not only allowed music to be premiered in video games but also to include actual musicians as playable characters. Among the notable examples of this has been MTV Games' *The Beatles: Rock Band* (2009) and the licensed Def Jam series of games, which features popular musicians from the Def Jam label in addition to their music (Marr, 2005). For those less inclined to build around music, the industry has begun to factor nostalgia into the mix as well, including re-releasing 'classic' video games such as *Pac-Man* and *Missile Command* (1980) to target gamers with more retro tastes (Schiesel, 2005a). One company reissuing older games says that the audience for classic reissues is made up primarily of adults between thirty and fifty-nine. Its estimate suggests that more than 70 per cent of its audience fits this demographic (Tran, 2002).

It seems logical that much of the recent success of video games owes itself to the industry's recognition of the true diversity possible in players. These differences are reflected throughout the industry from software production to a number of surprising responses via hardware production as well. Software, however, has been the most obvious means of attracting a variety of audiences.

THE SOFTWARE SECTOR

As has already been seen, software currently provides the majority of video game revenues. But because the software sector relies on innovation in hardware, adoption of a three- to five-year-production cycle for software has radically altered this aspect of the industry. Increasing graphics capability, now approaching movie quality, has pushed up development time and dramatically increased the cost of production (Snider, 2004). Rising production costs, as well as marketing costs, have consolidated power in the hands of a few large software publishers, forcing smaller companies – or developers – to fall into line (Dyer-Witheford, 2002). One exception has been the rise of small, casual game developers like PopCap and Zynga. Such companies have introduced an interesting tension into game production: should they focus on triple-A games, which one estimate suggests might require as much as $60 million to develop, or on casual games, which may run to hundreds of thousands of dollars (Deagon, 2010).

Companies working in software development fall into two categories: developers and publishers. Developers tend to be smaller groups, often as small as twelve to twenty individuals, who design specific elements. Kerr (2006b) offers a useful categorisation of developers. First-party developers are development firms owned and integrated into a publishing company; second-party developers create games based on concepts developed by or licensed through publishers; finally, third-party developers create their own projects that they then attempt to sell to publishers (Kerr, 2006b). Market concentration – particularly via integration – has meant, however, that third-party developers have had an increasingly difficult time getting original games into the market because licensed properties and franchise games are seen as less risky (Van Slyke, 2008). During successful years, it is not unusual to see large numbers of development studios purchased by publishers, which are sold, merged or closed during lean economic times. Similarly, when outside media companies attempt to move into the industry, it is typically through purchase or establishment of development studios. Publishers tend to be larger concerns affiliated with multiple developers, often referred to as studios. Studios tend to focus on particular types of games and, as will be discussed in Chapter 5, tend to work via individual contracts with publishers. Publishers, in turn, typically contract with hardware developers. It is not unheard of for hardware developers to vertically integrate, purchasing publishers and studios of their own. Both Microsoft and Sony are examples of this. For publishers, market concentration is considerably higher than for developers. Figure 2.6 details the concentration levels in publishing in 2008. With three companies making up more than 50 per cent of the market, and the industry's big four making up 44 per cent, entry into software publishing is difficult, and the business seems to have stabilised around a core group of publishers. One of the largest

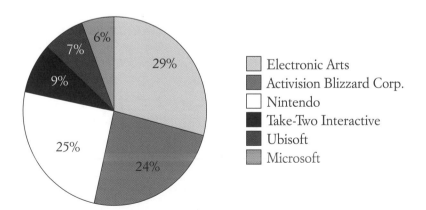

Figure 2.6 Top Video Game Publishers Market Share, 2008
Source: Market Share Reporter (2009b).

publishers in the industry is Electronic Arts, which has held its lead by developing software across platforms from its earliest days.

CASE STUDY: ELECTRONIC ARTS

Electronic Arts (EA) has been aggressive in its acquisition of studios and cautious in its support of particular platforms. Based in Redwood City, California, EA is the world's largest video game publisher. Table 2.2 provides information on Electronic Arts' facilities and key franchises. Founded in 1982, the company has seen the industry rise and fall. In its early years, it produced both productivity and video game software (People, 1985; Pitta, 1990). By the 1990s, its focus had shifted to video games, modelling itself on the film industry (Brandt, 1987). Rather than seek distributors, EA began to forge ties with retailers to secure shelf

Year formed	**1982**
Headquarters	**Redwood City, California, US**
2012 Sales (US$m)	**4,143.0**
2012 Employees	**9,200**
Industry sector	**Software publisher**
Key software franchises	*Command and Conquer*
	Def Jam
	FIFA
	Harry Potter
	James Bond, 007
	Jane's Combat Simulations
	The Lord of the Rings
	Madden NFL
	Medal of Honor
	MVP Baseball
	NASCAR
	NBA Live
	NBA Street
	NCAA Football
	Need for Speed
	NHL
	PGA Tour
	Rock Band
	SimCity
	SSX
	The Sims
	Tiger Woods
	UEFA
	Ultima Online
	Wing Commander

Table 2.2 Corporate Profile of Electronic Arts, Inc., 2012

Table 2.2 *continued*

Studios owned	Criterion Software – Guildford, UK
	Digital Illusions CE – Stockholm, Sweden
	EA Black Box – Vancouver, British Columbia, Canada
	EA Canada – Burnaby, British Columbia, Canada
	EA China – Shanghai, China
	EA Los Angeles – Los Angeles, California, US
	EA Montreal – Montreal, Quebec, Canada
	EA Casual Entertainment
	EA Mythic – Fairfax, Virginia, US
	EA Korea – Seoul, South Korea
	EA Byrnest – Mount Sinai, New York, US
	EA Redwood Shores – Redwood City, California, US
	EA Singapore
	EA UK – Guildford, UK
	Maxis – Emeryville, California, US
	EA Phenomic – Ingelheim, Germany
	EA Tiburon – Maitland, Florida, US
	EA Salt Lake – Bountiful, Utah, US
	EIS (European Integration Studio) – Madrid, Spain
	EA Mobile – Bucharest, Romania
	EA Mobile – Hyderabad, India
	EA Atlanta – Atlanta, Georgia, US
	BioWare – Edmonton, Alberta Canada and Austin, Texas, US
	Pandemic Studios – Los Angeles, California, US and Brisbane, Queensland, Australia

Sources: Hoover's (2012); Mergent (2008).

space in the mid-1980s as a way to insulate itself from big changes in the market (Business Week, 1985).

This early move allowed EA to make the leap from developer to publisher within a decade. Its founder and former president, Trip Hawkins, is said to have formed EA with an eye on the Hollywood studio system, facilitating ideas generated from a wide variety of sources but with the company in ultimate control of which games went out (Pitta, 1990). In 1990, EA published 350 titles, but only around 100 were developed in-house; the rest were licensed from other developers (Pitta, 1990). One major reason for this was the cost of developing a title across platforms. According to Hawkins, 'for every dollar it costs to develop a new title, it costs another fifty cents to move it to a new platform' (Pitta, 1990). Because EA did not produce its own platforms, but rather licensed across them, it had to adapt to the rapid number of platforms on the market.

In the early 1990s, it refocused, making video games its priority. Unlike the other three industry leaders, it has focused solely on game development.

However, because it is not tied to any single platform, it has capitalized on the competition between the three console giants as well as that for personal computers (EA, 2005; Hoover's, 2002, 2005).

EA has used its dominance to take the lead in licensing and merchandising deals, a practice similar to the Hollywood studio system. It has long-term arrangements with the major sports leagues NBA, NFL, Major League Baseball, the Collegiate Licensing Corporation, FIFA, cable network ESPN and a number of Hollywood studios. It has also made hit games based on the 'James Bond,' 'Lord of the Rings' and 'Harry Potter' franchises. The company's board of directors has been tied to a number of major concerns and media outlets including CBS, Warner and Polygram; the Omnicom Group, a leader in advertising; Babbages, Inc., a software retailer; General Electric, AOL and Young and Rubicam advertising (EA, 2005; Hoover's, 2002, 2005). The company has nine different facilities in the US and Canada, as well as the UK, Germany, Sweden, Switzerland, Romania, India, China, South Korea and Singapore (EA, 2010). EA has been so successful that it not only leads video game software publishers but is in the top ten of all software publishers in the world (Market Share Reporter, 2008a).

Software Development

In the US alone, more than 240 million video games were sold in 2006. That amount equates to almost two games for every household in that year alone. Almost 85 per cent of those sales were console games (ESA, 2007). Figure 2.7 provides a comparison of software sales by sector in the US from 2003 to 2009. It clearly indicates a steady growth in the sales of software in spite of economic downturns. It's also significant that the sale of portable games is not broken down in these figures, in spite of its growing importance to the industry.

A closer look at the bestselling games affords a snapshot of power in the industry. Between October and December 2004, for example, two games sold more than 3.3 million units in the Christmas season – Microsoft's *Halo 2* and Rockstar's *Grand Theft Auto: San Andreas* (Richtel, 2004). Other estimates suggested that *Halo 2* sold more than 5 million units by the end of 2004, an impressive feat as it was only available for the Xbox console (Fritz, 2004). Another game, Blizzard's *World of Warcraft*, sold more than 240,000 units in its first twenty-four hours of combined sales in the US, Australia and New Zealand. By the end of 2004, *World of Warcraft* had sold 700,000 copies (Schiesel, 2005a). Estimates suggest that by 2009, *World of Warcraft* alone accounted for $1 billion in revenue for Blizzard (Sarno, 2010). These numbers are impressive because in the video game industry only 5 per cent of all games sell more than a million units. Industry giant Electronic Arts had twenty-seven titles break the 1 million mark in 2004 (Grover *et al.*, 2005).

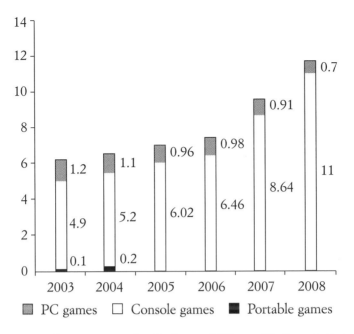

Figure 2.7 US Software Revenue by Platform in Millions of Dollars, 2003–9
Sources: ESA (2005a, 2006, 2007, 2010).

Part of the industry's success owes to its cultivation of franchises. In the early years, many of these franchises came from outside, but as the industry matured in the 1990s, a number of companies began to develop their own. Microsoft's *Halo* franchise, which originated in 2001, has sold more than 12.8 million units and earned the company more than $600 million in revenue (Brodesser and Fritz, 2005). It was so highly anticipated that prior to its release, the second title in the series, *Halo 2* (2004), had sold more than 1.5 million units in the US alone (Oser, 2004). It has sold more than 6.3 million units since its release (Marriott, 2005a). Combined, *Halo 2* and *Grand Theft Auto: San Andreas* (2004) sold almost as many units as all other top ten selling titles combined (Newman, 2005). Tables 2.3, 2.4, 2.5, and 2.6 display the bestselling software for Europe, Japan, the US and Australia respectively. Whenever possible, the number of units and the platform for each game is included, but – particularly in the European case – this information is not always readily available and because many titles are released for multiple platforms, an educated guess is not possible. What is most striking is the number of titles that are common across regions. The data also indicate a clear concentration in favour of a very few major publishers and, perhaps most strikingly, the sales patterns in terms of consoles. In the time period represented, it's clear that the Nintendo Wii was the dominant platform globally, but this dominance didn't translate across regions. The US

Game title	Platform	Publisher
Dr. Kawashima's Brain Training	Data unclear	Nintendo
FIFA 2008	Data unclear	Electronic Arts
Wii Play	Wii	Nintendo
Pro Evolution Soccer 2008	Data unclear	Konami
New Super Mario Bros.	Wii	Nintendo
Need for Speed: ProStreet	Data unclear	Electronic Arts
Assassin's Creed	Data unclear	Ubisoft
Call of Duty 4: Modern Warfare	Data unclear	Activision
Big Brain Academy	Data unclear	Nintendo
The Simpsons Game	Data unclear	Electronic Arts

Table 2.3 Top Ten Video Game Titles in Europe, 2007
Source: Gamasutra (2008).

Game title	Platform	Publisher	Units sold in millions
Wii Sports	Wii	Nintendo	1.98
Hajimete no Wii	Wii	Nintendo	1.54
Monster Hunter Portable 2nd	PSP	Capcom	1.50
Pokémon Fushigi no Dungeon: Yami no Tankentai	DS	Nintendo	1.32
Mario Party DS	DS	Nintendo	1.26
New Super Mario Bros.	DS	Nintendo	1.21
Mario Party 8	Wii	Nintendo	1.11
Pocket Monsters Diamond/Pearl	DS	Nintendo	1.10
Touhoku Daigaku Mirai Kagaku Gijutsu Kyoudou Kenkyuu Center Kawashima Ryuuta Kyouju Kanshuu Motto Nou wo Kitaeru Otona no DS Training	DS	Nintendo	1.08
Dragon Quest IV: Michibikareshi Monotachi	DS	Square Enix	1.05

Table 2.4 Top Ten Video Game Titles in Japan, 2007
Source: VGChartz (2008a).

showed the widest range of consoles used, while Japan indicated a clear preference for handheld platforms. Finally, while the Australian data are more dated,
they seem to show the importance of the PC platform, particularly for regions
where later console generations have not penetrated the market.

Like the hardware sectors, software is an increasingly international affair. In
1998, for example, Japanese software made up almost 50 per cent of the software
sold in the US market, and though this number dropped to 29 per cent by 2004,

Game title	Platform	Publisher	Units sold in millions
Wii Sports	Wii	Nintendo	7.40
Halo 3	Xbox 360	Microsoft	4.99
Pokémon Diamond/Pearl	DS	Nintendo	4.40
Wii Play	Wii	Nintendo	4.31
Call of Duty 4: Modern Warfare	Xbox 360	Activision	2.77
Guitar Hero III: Legends of Rock	PS2	Activision	2.50
Super Mario Galaxy	Wii	Nintendo	2.50
Forza Motorsport 2	Xbox 360	Microsoft	2.36
Nintendogs	DS	Nintendo	2.18
Mario Party 8	Wii	Nintendo	1.98

Table 2.5 Top Ten Video Game Titles in the US, 2007
Source: VGChartz (2008a).

Game title	Platform	Publisher
Gran Turismo 4	PS2	SCE Australia
Grand Theft Auto: San Andreas	PS2	Take 2
Pokémon Emerald	Gameboy	Nintendo
Need for Speed: Underground 2	PS2	Electronic Arts
Simpson's Hit & Run Platinum	PS2	Vivendi
Super Mario 64	DS	Nintendo
The Sims 2: University	PC	Electronic Arts
Burnout 3	PS2	Electronic Arts
World of Warcraft	PC	Vivendi
Grand Theft Auto: San Andreas	Xbox	Take 2
Need for Speed: Underground Platinum	PS2	Electronic Arts
Ratchet & Crank 2 Platinum	PS2	SCE Australia

Table 2.6 Top Ten Video Game Titles in Australia, 2005
Source: IEAA (2005).

this is due to increasing moves by European, Canadian, Australian and even Chinese software designers (Croal and Itoi, 2004). In China, video games have become increasingly popular, in spite of not always being sanctioned by the Chinese government. Games sales experienced 56 per cent growth between 1998 and 2003, a figure representing only legitimate software sales in a country where pirated games are the norm (Euromonitor International, 2005). Table 2.7 details the top software publishers for 2007, including how many games each released and how many studios they own. As the table indicates, there is a tremendous amount of size variance among the companies, with the three major hardware producing concerns, Sony, Microsoft and Nintendo, having resources

Publisher and year formed	Revenue in millions of dollars, 2007	Number of games released for 2007, 2006, 2005, 2004				Number of internal studios 2007, 2006, 2005			Number of employees in 2007
Nintendo (1889)	8,188.4	32	39	20	36	5	5	7	Not reported
Electronic Arts (1982)	3,665	116	139	126	96	15	15	10	9,000
Activision (1979)	2,898.1	79	45	76	ND	11	11	10	2,125
Ubisoft (1986)	660.6	86	67	65	35	14	14	13	3,441
THQ (1989)	1,026.9	65	76	94	69	16	16	11	2,000
Take-Two Interactive (1993)	981.8	56	57	42	28	12	12	13	1,900
Sega of America (1952/1975)	4,704.8	31	39	75	30	7	7	7	6,416
Sony Computer Entertainment (1993)	70,513.4	28	43	41	36	16	16	14	163,000
Microsoft Game Studios (1975)	51,122	15	12	20	32	6	6	6	79,000
Sci/Eidos (1990)	325.2	47	25	35	ND	4	4	5	900
Square Enix (1975/1996)	1,058.5	12	10	19	11	4	4	5	3,050
Namco Bandai (1955/1950)	3,833.8	38	58	35	25	6	6	2	3,096
Vivendi Games (2000)	26,444	35	26	43	78	9	9	6	37,014
Capcom (1979)	No data available	15	35	39	ND	6	6	4	1,175
Konami (1973)	2,374.2	42	57	78	23	4	4	4	12,046
NCSoft (1997)	352	4	ND	ND	ND	3	3	ND	3,000
Disney Interactive Studios (1994)	35,510	34	21	ND	ND	5	2	ND	137,000
Atlus USA (1991)	17	18	ND	ND	ND	ND	ND	ND	16
LucasArts (1982)	Not reported/ private	12	6	9	ND	1	1	1	1500
Midway (1988)	157.2	25	38	22	31	6	6	6	900

Table 2.7 *Game Developer*'s Top Twenty Publishers, 2007

Notes: When publisher is a subsidiary, sales and employees are reported for the entire company.

ND = No data reported in the rankings for that year.

Sources: Donovan (2004, 2005); Hoover's (2008); Wilson (2006, 2007); and industry sources.

well beyond almost any other on the list. EA, which was the fourth largest company in the industry, is considerably smaller, though it owns a number of studios and releases more games than any other company on the list. The chart also demonstrates some of the volatility that goes with working in a development studio, even one owned by a publisher. Between 2005 and 2007 a number of publishers sold off some of their studios or merged them with other development teams. That volatility contributes to some worker dissatisfaction. As will be seen in Chapter 5, the labour situation for those in software development can be particularly difficult because workers and studios are often contracted for only one game at a time, forcing them into a constant search for their next position.

Software development also owes to views of its past games as well as to the ties between publisher and hardware manufacturer; it is not uncommon that a hardware manufacturer will license with a publisher to create games based on their systems architecture rather than allow open source development (Dyer-Witheford, 2002). Such ties are crucial as the time and budget needed to develop games have risen drastically. In 1983, the hit game *Frogger* cost roughly $5,000 to develop, but by the year 2000, a top game could cost more than $1 million and take up to eighteen months to develop (Delaney, 2003; Zito, 2000). Development for major games can now take between eighteen and twenty-four months (Levine, 2005c). Termed 'Triple-A' titles, bestselling games can also cost between $10 to 20 million to develop (Richtel, 2005b). Development for the most recent consoles – Microsoft's Xbox 360, Sony's PlayStation 3 and Nintendo's Wii – was estimated to raise the overall cost by an additional $3 to 6 million per game (Dee, 2005; Taub, 2005). Others have suggested development costs might triple (Gentile, 2005b). As Chapter 5 will demonstrate, the labour required to create games will drastically increase, involving ever larger teams of developers (Guth *et al.*, 2005; Guth and Divorack, 2005).

The power of the publishers in the industry is considerable. Microsoft, which produces both its console and a considerable amount of software, has begun to cut back production of its own in-house games in an attempt to try to connect with innovative software developers. It has issued special developer packs and signed deals with well-known Japanese developers in an attempt to increase profitability (Guth *et al.*, 2005). One of Microsoft's biggest moves was to sign a deal with Hironobi Sakaguchi, the creator of the *Final Fantasy* series (1987), to design games for the next generation Xbox (AP, 2005b). These were first made available in Japan in an attempt to boost Microsoft's sales in a market where it has historically performed badly. Having a marketable name – either a game creator, development team or even publisher – can make the difference in a game's success. Figure 2.8 provides a historical comparison of concentration of top ten bestselling titles by publisher in 2001 to 2004 and 2010 to 2013. It should be

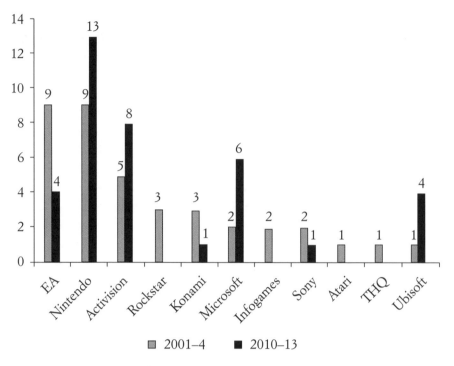

Figure 2.8 Total Number of Top Ten Titles by Publisher, 2001–4 and 2010–13
Sources: NPD (2002, 2003, 2004, 2005); VGChartz (2013a, 2013b, 2013c, 2013d).

noted that the figures from 2001 to 2004 only represent sales in the US because global figures were not available while the 2010 to 2013 figures represent global sales. Since the US dominated sales however, its figures can serve to suggest the global ones. Such numbers can't indicate complete dominance, particularly because of the number of titles sold each year, but the data does suggest a pattern of heavy concentration in the early period extended through hardware control. While two of the big four companies – EA and Nintendo – again show their dominance of the market, both Sony and Microsoft have had a tougher time. Their control of hardware platforms has proved crucial for them, and for keeping smaller publishers from establishing any lasting control of the market. However, by the later period, the dominance of publishing had clearly shifted. Hardware control still mattered, as Wii helped Nintendo dominate much of the later period, but several companies – the US-based Activision and French firm Ubisoft, in particular – both experienced growth rooted in key software franchises and licensing deals.

Licensing tends to happen between the publisher and hardware manufacturers, while most developers in the industry negotiate directly with publishers themselves (Williams, 2002). This has allowed most publishers to develop and license games across platforms (Dyer-Witheford, 2002). Often licensing happens on a year-to-year basis with individual publishers licensing deals covering individual pieces of hardware (for example, the Sony PlayStation 3). This forces publishers to negotiate with hardware manufacturers for each platform. In many cases, these deals roll over automatically from year to year unless there is a substantial change on the part of one party or other (Atari, 2008). As Chapter 4 will detail, publishers are also most commonly the ones to negotiate cross-industry licensing, particularly with the global film and recorded music industries (Fritz, 2005c, 2005d). One of the most notable examples of this is Disney's recent acquisition of software developer Avalanche as well as its creation of a studio in Vancouver (Marr, 2005). The Vancouver studio lured workers away from the music business. This has resulted in a number of major media companies' on-again-off-again flirtation with video games (Fritz, 2005c). It has been Disney's hope that, rather than licensing its products to video game publishers, it will ultimately become one itself, making 80 to 90 per cent of its game revenues in-house (Marr, 2005). Like feature films, revenue within the video game industry has expanded beyond basic sales. Software publishers have taken a page from the Hollywood book, working to license their products into as many other formats as possible, including movies, books, TV shows and merchandising (Bloom, 2001; Cooper and Brown, 2002; Elkin, 2003; Moledina, 2004a; Wingfield and Marr, 2005). Licensing deals are increasingly competitive and costly. In early 2005, EA spent more than $800 million to lock up an exclusive fifteen-year deal with ESPN (Grover et al., 2005).

CASE STUDY: ACTIVISION-BLIZZARD

Santa Monica-based outfit Activision-Blizzard is one of the oldest software publishers in the industry. Founded in 1979, the company published a number of games for the earliest consoles including the hit game *Pitfall!* (1982) for the Atari 2600. The company started as a small software developer, providing games for Atari and other early game systems, but its profitability eventually allowed it to become a major publisher in its own right, rivaling EA. Table 2.8 breaks down the company's major franchises and studios. Its string of successes continued, with games like *Doom*, *Underground* (2003) and *Call of Duty* (2003). The company has also managed a number of impressive licensing agreements for major media products. It has published games based on LucasArts' 'Star Wars' franchise, Sony's 'Spider-Man' movies and DreamWorks' 'Shrek' (Activision, 2004; Hoover's, 2005). In 2008, media and communications conglomerate Vivendi bought a majority stake in the firm valued at $9.8 billion, and combined the company with its Blizzard Entertainment division to form the new Activision-

Year formed	1979
Headquarters	**Santa Monica, California, US**
2011 Sales (US$m)	**4,755**
2011 Employees	**7,300**
Industry sector	**Software publisher**
Key software franchises	*Guitar Hero*
	Tony Hawk
	Call of Duty
	Shrek
	Spider-Man
	X-Men
	Transformers
	James Bond
	Crash Bandicoot
	Spyro
	Starcraft
	Warcraft
	Diablo
Studios owned	Beenox – Quebec City, Quebec, Canada
	Blizzard Entertainment – Irvine, California, US
	Bizarre Creations – Liverpool, UK
	High Moon Studios – San Diego, California, US
	Infinity Ward – Los Angeles, California, US
	Luxoflux – Santa Monica, California, US
	Massive Entertainment – Malmö, Sweden
	Neversoft – Los Angeles, California, US
	Radical Entertainment – Vancouver, British Columbia, Canada
	Raven Software – Madison, Wisconsin, US
	RedOctane – Mountain View, California, US
	Shaba Games – San Francisco, California, US
	Sierra Entertainment – Los Angeles, California, US
	Sierra Online – Los Angeles, California, US
	Toys for Bob – Novato, California, US
	Treyarch – Santa Monica, California, US
	Vicarious Visions – Albany, New York, US
	Z-Axis – Foster City, California, US

Table 2.8 Corporate Profile of Activision-Blizzard, Inc., 2012
Sources: Hoover's (2012); Mergent (2008).

Blizzard (Hoover's, 2008), which has studios in the US, Canada, the UK, France, Germany, Ireland, Italy, Sweden, Spain, Norway, Denmark, the Netherlands, Romania, Australia, Chile, India, Japan, China, Taiwan and South Korea (Activision, 2008). The board has ties to Proctor and Gamble; a number of investment banks; Four Kids, Inc., a company focused on the licensing and merchandising of children's products; Universal Music Group; Canal+; Avon; and Random House (Activision, 2004, 2007; Hoover's, 2005).

Activision was the leading third-party publisher in 2007, behind Electronic Arts, and delivered fifteen years of revenue growth. In 2007, it was ranked third in publishers in Europe. It also completed its acquisition of RedOctane Studios in 2007 as well as of DemonWare, the leading provider of networking technology for consoles and PCs. The company produces software for PCs and all of the major consoles, but a majority of its revenues in 2007 came from games for Sony's PlayStation 2 (Activision, 2007). As seen earlier, Activision has built upon its successes, with an increasing number of bestselling game titles each year, enabling it to rival Electronic Arts as the leading game publisher globally.

Unlike many in the software sector, Activision has forged its own path, emphasising franchise development, merchandising and global expansion. In 2009, Activision CEO Bobby Kotick committed to these processes rather than follow the sector's trend of purchasing developers focused on creating social games for platforms such as Facebook (Edwards, 2012). Indeed, much of the company's recent success has come from its profitable managing of franchise video games. Its two biggest selling franchises are the *Call of Duty* games and *World of Warcraft*. Estimates suggest that various *Call of Duty* games earned US $37 billion through 2012 while *World of Warcraft* netted approximately US $1 billion annually (Economist, 2011).

The *Call of Duty* franchise is illustrative of the realities of the software sector in the modern video game industry. The first *Call of Duty* game was released in 2003, developed by Infinity Games and published by Activision. By 2012 a total of eight games had been released, three of which ranked among the bestselling of all time (Grossman and Narcisse, 2011). The franchise has proved so popular that the company has been known to compare it to the *Star Wars* franchise or the NFL in terms of its profitability, as evidenced by the growth in sales from one game to the next. *Call of Duty: Modern Warfare 2* launched in November 2009, earned US $310 million in its first day of sales, making it the most successful launch of any media product in history. By July 2010, the game had earned more than US $1 billion globally (Grossman and Narcisse, 2011). Such a figure is impressive, until compared with the 2011 release *Call of Duty: Modern Warfare 3*, which took only sixteen days to earn US $1 billion globally (Satariano, 2010; Suellentrop, 2012). The company has emphasised its aim of releasing a new game in the franchise every two years, which necessitated leveraging a broader range of its resources. The first *Call of Duty* game was developed by the Infinity Ward studio, but the eighth game in the franchise is estimated to have had between 400–500 employees from development studios including Infinity Ward, Sledgehammer Games and Treyarch (Business Wire, 2004; Satariano,

2010). According to the company, this has been necessitated by advances in technology, but marketing has also become costly, with one estimate suggesting that *Call of Duty: Modern Warfare 3* had a marketing budget of more than US $100 million (Suellentrop, 2012). Taking advantage of the capabilities of the seventh generation of consoles – the Xbox 360, PlayStation 3 and Wii – resulted in skyrocketing development costs, sometimes tripling those of the previous generation (Satariano, 2010). Estimates in 2012 suggested that games with similar development costs needed to sell at least 2 million copies in order to turn a profit; such costs mean that only those developers with bestselling franchises stand a chance at profitability (Satariano, 2010).

In addition, the company has been successful in its exploration of alternative revenue streams. With the 2011 release of *Skylanders: Spyro's Adventure*, the firm had a franchise that relied not just on the sales of games but also on a line of associated toys, both developed by Toys for Bob, a studio owned by Activision (Activision, 2008; Edwards, 2012). Games come bundled with software, trading cards and three action figures; consumers can then purchase additional merchandise as well. While the sales of the games are impressive – more than 32 million copies had been sold by March 2012 – the merchandise figures cannot be ignored. In the first six months of 2012, it is estimated that more than US $99 million in *Skylanders* toys were sold, more than the combined sales of Hasbro's G.I. Joe and Transformer toys that were estimated to have earned US $54 million in the same time frame (Edwards, 2012).

Activision has also concentrated on expanding the global market. Perhaps the best example of this can be seen in the company's partnerships with two major Chinese internet providers – NetEase and Tencent – in order to release versions of its games in China. Its first partnership with NetEase allowed the firm to create regionalised versions –with content specifically targeted to the region it was to be sold in – of *World of Warcraft* and *StarCraft* (1998) that the Chinese government would approve for sale (China Economic Review, 2012; Moscaritolo, 2012). Online gaming in China has become a major revenue stream for those able to get their products into the market. In 2009, online game sales in China were estimated to have jumped approximately 30 per cent, to roughly US $4 billion (China Economic Review, 2010). NetEase's revenues from online games made up a sizeable portion of that market. In 2010, the company is estimated to have earned US $189 million in its third quarter alone (China Economic Review, 2010). With the introduction of the *Call of Duty* franchise in partnership with Tencent, Activision positioned itself for entry in one of the most anticipated markets globally (Moscaritolo, 2012).

Activision's rise to success has not been without controversy, however. Shortly after the release of *Call of Duty: Modern Warfare 2*, the company fired the founders of the Infinity Ward studio who created the franchise. They then signed deals to develop content for EA and sued Activision. Activision emphasised delays to the game in its arguments, while the employees cited the harsh work demands required to meet production deadlines (Satariano, 2010). While the lawsuit was ultimately settled out of court in 2012, it is noteworthy for a few reasons. First, because it establishes a pattern of litigation on Activision's part, and such a pattern is probably emblematic of the industry trends at large. The company was sued by Harmonix, developer of the *Guitar Hero* (2005) franchise; and Activision sued developer Double Fine Productions in an attempt to stop the release of the game *Brutal Legend* (2009) (Satariano, 2010). While both those lawsuits were either dropped or settled, it points to an increasing reliance on litigation as well as highlighting the difficult working conditions faced by many in the software sector, a point taken up in Chapter 5.

Software Genres

Currently, most video game revenue comes from game sales. In addition, the industry has begun to gather revenue from subscriptions to online games, typically from some form of Massive MultiPlayer Online Role-Playing Game; in these, players are often willing to pay monthly subscription fees on top of the initial purchase of the game (ESA, 2004, 2005a; IDSA, 2003; Irvine, 2004). Activision's *World of Warcraft* is an excellent example of this genre. These games let players create characters, referred to as 'avatars' in order to navigate through elaborate, expansive worlds while interacting with other players. This results in continued work for designers who must keep the world updated and working (Schiesel, 2005a). Such games are risky and costly to develop; if they don't prove popular, it can spell big losses for both the developer and the publisher (Delaney, 2003; Levine, 2005a; Slagle, 2005b).

One way some developers have tried to work around this has been to re-release former bestsellers for new platforms; this has cut the development cost and the risk of gambling on an unknown quantity. French software developer Infogames has begun to reissue old Atari games for the latest platforms. The cost of developing these older games has been a little over $200,000 (Tran, 2002).

A second work-around has been the invention of casual games. These are targeted at a mass audience with limited time to play. They tend to be fairly simple in design, so that players require only minimal knowledge, and to feature gameplay allowing for either interruption and continuation or for repeat play. They tend to use alternative modes of distribution to PC and console games, such as direct download or embedding in another media form, such as a website. Though their sales started small – representing only about $250 million in sales for the industry in 2004 – they are becoming increasingly popular

(Marriott, 2005b). One 2006 estimate suggested that casual games could grow to be an $8 billion dollar industry (Moltenberry, 2006). One example is *Diner Dash* (2003), which has been sold exclusively on the Internet. It has become one of the most downloaded games from a number of sites including Yahoo Games, Real Arcade and Shockwave.com. By the beginning of 2005, it had sold more than 50,000 units at $20 per game, and it continued to sell 1,000 copies a day for much of that year (Marriott, 2005b).

In some cases, these games are tied to social networking sites like Facebook and MySpace, in which case they are referred to as 'social games,' a subset of the casual games market. Casual games have the benefit of being extremely affordable to develop, even as they earn high profits (Deagon, 2010). Facebook surpassed 200 million users in 2009 and had amassed more than 1 billion by 2012 (Learmonth and Klaassen, 2009; Thier, 2012). Such sites have become big business. Games like *Farmville* (2009), *Mafia Wars* (2009), and *Zynga Poker* (2007) have scored big for developer Zynga, which was predicted to earn more than $1 billion from the sale of virtual goods in 2010 (Patel, 2010). Estimates suggest that in 2010, there will be $1.6 billion in virtual good sales in the US and £50 million in the UK (Smith, 2010). One of the key benefits of the casual game is that it draws an audience that is of growing importance to the industry: female gamers, particularly older female players (Abraham *et al.*, 2010; Wayne 2010). One estimate by PopCap games suggests that the most common demographic for a casual gamer is a woman in her forties. In fact, the company indicates that more than 70 per cent of the visitors to their website are female, and more than 75 per cent of them are over thirty-five (Moltenberry, 2006). Moreover, these games draw repeat players, with 90 per cent of players in one survey indicating that they play at least once a day, and a majority of those – 62 per cent – indicating that they play multiple times in one day. These players are also highly loyal, particularly those who pay for virtual goods. According to the same, of those who spent more than $25 on virtual goods, only about 15 per cent indicated that they had done so in more than one game (Macmillan, 2010). Zynga estimates only 1.2 million of their 60 million unique visitors to *Farmville* purchase goods, but because the games typically cost only 'a few hundred thousand dollars to develop', the profits can be considerable (Deagon, 2010). While Zynga's fortunes have faded, particularly following its initial public offering, the profitability of social games still presents a disruption to the pattern of game development and profitability that existed earlier in the same decade (Flare, 2012).

One reason that casual games are gaining more attention is the ease with which they can be transferred to mobile platforms, particularly cell phones. In 2004, software for mobile games saw revenues of almost $204 million in the US (AP,

2005d). Globally, however, one study suggests mobile games have sold more than 42.3 million units, giving the industry sales of $1 billion (Gamedaily.biz, 2005). This has prompted a number of Internet companies, including Internet giant Yahoo!, to enter the mobile games business (AP, 2005d). Mobile games are also significant because customers can be billed a monthly fee per game, typically between $1.50 and $3 per month, with somewhere between 25 and 35 per cent of the profit going to cell phone carriers (Richtel, 2005c).

Other companies, like PopCap Games have worked to reduce cost by seeking low-cost distribution channels. PopCap has focused on developing games for direct download, a feature that has helped the company as digital distribution has taken hold with platforms like the Xbox 360, PlayStation 3 and mobile phones. In addition, the firm has sold its games at lower prices in major retailers, opening a second avenue into the marketplace. The Seattle-based company, founded by three friends in 2000, has focused on smaller game production across a variety of platforms, including PC, social networking sites and smart phones (Cook, 2006; Seitz, 2010; Strong, 2003). One of its most popular games, *Bejeweled* (2000), has had more than 200 million downloads and its website has more than 7 million unique visitors per month (Gibbs, 2007). With an initial team of just the three friends, the company emphasised simplicity, trying to make sure its games could be played on '1998-era computers with 4.0 browsers and 56K modems' (Strong, 2003). PopCap was so profitable that it was purchased by EA in 2011 (De La Merced, 2011). While its revenues were not released prior to its purchase, president David Roberts indicates that PopCap had been profitable from its very beginning (Seitz, 2010). The company went through a variety of business models in rapid succession, including advertiser-supported, only to settle on a shareware model that catapulted revenues (Moltenberry, 2006). The company claims that within two years of its founding, its sales were in the millions of dollars (Strong, 2003).

The popularity of casual games and their ability to generate sales not necessarily tied to a particular console has attracted attention from more traditional video game companies, venture capitalists and outside media investors (Cook, 2006). In November 2009, Electronic Arts bought casual game developer Playfish for $300 million. In the same year, Google purchased rival developer Slide for $228 million, and Disney snapped up Playdom for $563 million (Deagon, 2010). The success of such companies has drawn users away from established game sites like Pogo and AOL Games, which both lost more than 15 per cent of their audience from January 2009 to the start of 2010. Likewise, Yahoo Games lost nearly 30 per cent from January 2009 to January 2010. MSN Games has suffered from the casual game boom, as well, losing more 3 million unique users from 2006 to 2010 (Shields, 2010).

This question of how to most affordably generate content is one of the areas that has bound video games to so many other cultural industries. Because they are relatively new on the scene, while film, television, sports and recorded music have decades of history and content available, borrowing content has become big business (Nichols, 2008). Working with proven media franchises can reduce the risk in creating such costly products. Electronic Arts has been particularly good at this, with recent releases based on movies *The Godfather* (1972) and James Bond classic *From Russia with Love* (1963), while Rockstar Games (owned by Take-Two Interactive) has created a game based on the cult classic *The Warriors* (1979). Chapter 4 discusses the relationship between video games and film in more detail.

The industry divides itself into a very particular series of genres, which shift slightly depending upon the platform considered. This division has served two major purposes: first, to help classify games in such a way that producers know which audience segments they are designing for while those audiences know what to expect; and second, to avoid regulation wherever possible (Ruggill, 2004). Figures 2.9 and 2.10 break down historical sales by genre for PCs and consoles in the US, while Figure 2.11 displays similar data for Australia. One problem revealed by this data is the slippery notion of genre itself. In the US case, the industry trade group, the Entertainment Software Association (ESA),

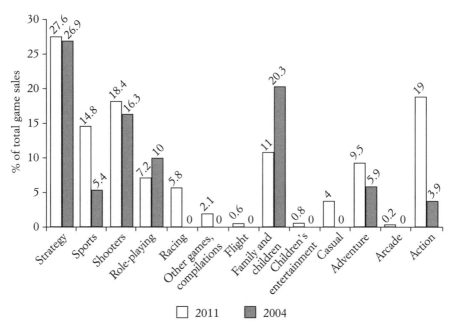

Figure 2.9 Bestselling PC Games in US by Genre in 2004 and 2011
Source: ESA (2005a, 2010).

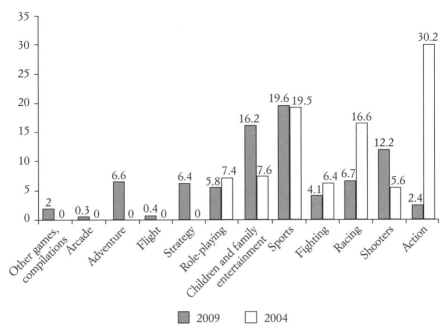

Figure 2.10 Bestselling Console Games in US by Genre in 2004 and 2009
Source: ESA (2005a, 2010).

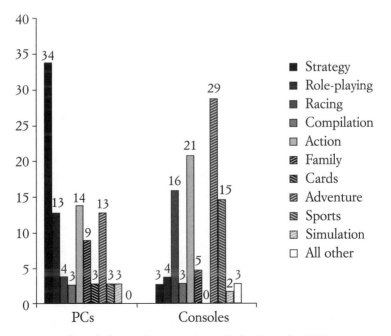

Figure 2.11 Bestselling Software Genres in Australia by Console, 2007
Source: IEAA (2007).

has created a system of genre classification, but this has been changed periodically, with new genres added or removed from the listing. This is seen in both
Figures 2.9 and 2.10, where the genre categories have changed between 2004
and 2011. Genre sales data are also difficult to obtain in the European and
Japanese contexts, though similar patterns seem likely in both cases. In spite of
this, however, some interesting trends do emerge. It is apparent that there is a
relationship between the success of a genre and the platform itself, as sales of
some genres vary depending upon which platform they've been released for.
Whether this is a function of the audiences associated with each platform or
some other feature or combination is unclear. This same pattern is seen in the
Australian market. For example, role-playing games are more popular in the PC
market. This is probably due to the fact that it is more common for PCs to be
connected to the Internet, though consoles have long had that capability and it
is almost a given for current models. Surprisingly, the genre that draws the most
revenue is not the action genre. Sports has become one of the most important
genres, with licensing deals between major leagues and organisations having
tremendous pull (AP, 2005a; Fritz, 2002, 2005a). Sports have helped propel
software leader EA into its dominant position by providing stable, consistent
sellers such as the Madden series of football games (Wahl, 2005). EA's licensing deals with the NFL, ESPN and others have helped to insulate the company,
even from setbacks such as shortages of hardware shipping (Flynn, 2005).
Statistics show, in fact, that EA was 'the fourth largest capitalized software
maker in the world behind Microsoft, Oracle, and SAP' (Lowenstein, 2005).

The last area of software development relies on the same premises as casual
games – cheap production, simple gameplay – but has focused on integrating
the educational and ideological potential of video games. Dubbed 'serious
games', these are often developed for political or training purposes (Wallace,
2004). Examples include *Howard Dean for Iowa* (2003), a game used in the Iowa
campaign for Howard Dean's presidency, and the VRPhobia project, which uses
games to treat phobia and other disorders via cognitive theory. Because these games
come from such varied sources, there has been little success in tracking development costs, but such games are likely to become increasingly common.
Another example, *America's Army*, produced by the US Army as a recruiting and
training tool, suggests that these serious games are one way in which video games
may be useful to the state (Nichols, 2009; Nieborg, 2009).

Regulation

While the video game industry operates on a global scale, regulation tends to
happen at the national, regional and sometimes even state levels. Significantly,
most regulation specifically targeted at video games has focused on software

ratings. In most cases, games are given ratings by some form of government board; the US is an exception to this, citing fears of censorship and its impacts on game sales, though little evidence has been given to support this fear. In the US, and to some extent in Canada, this has resulted in a ratings system controlled by the industry, reliant upon its breakdowns of content and measures of its appropriateness. As seen previously, this has resulted in a situation where the categorisation system itself can change with little or no explanation.

From the industry standpoint, the ideal situation would be much as it is in the US with most regulation happening within the industry with as little government involvement as possible. As Ruggill (2004) has noted, this is ideal both for sales and for staving off government regulation. However, in most areas of the world, games regulation involves considerably more government input, though the areas of focus differ. Content and distribution may be regulated based on decency standards, but it is increasingly apparent that governments also wish to support national industries.

In the US, video games, as with many media industries, have been largely unregulated by the government. In part, this may be due to a perception that they, as US District Judge Stephen Limbaugh has said, 'have no conveyance of ideas, expression, or anything else that could possibly amount to speech' (Morris, 2003). Ironically, this view has been advocated by the US military even as it has adapted video games as part of an ongoing public relations campaign. The military view is that games must be legislated because they are 'law enforcement training devices' (Kent, 2003). These views, however, seem to be in the minority, as the bulk of attempts at legislation of video games have come from state rather than federal government, with attempts to legislate against violent content and the sale of games in states like Illinois, Alabama, Missouri and Washington (AP, 2005c, 2005d).

Like the film and recorded music industries, most regulation of video games within the US has been handled by an industry-run board. The Entertainment Software Ratings Board (ESRB) has been active since the mid-1990s and has rated more than 10,000 video games (Felberbaum, 2005). Figure 2.12 provides a breakdown for sales by rating in the US and Canada. The ESRB has been active in maintaining ratings standards, modelling itself on the Motion Picture Association of America's (MPAA). It is worth noting that the ESRB is responsible for the ratings system in both the US and Canadian markets. At the heart of both systems is a series of basic age ratings markers applied to the products themselves as well as to advertising. Table 2.9 offers a comparison of ratings systems worldwide, showing considerable variance between them. Many include both age and content descriptors and most, with the exception of the ESRB's system in the US and Canada, are run by state-sanctioned agencies.

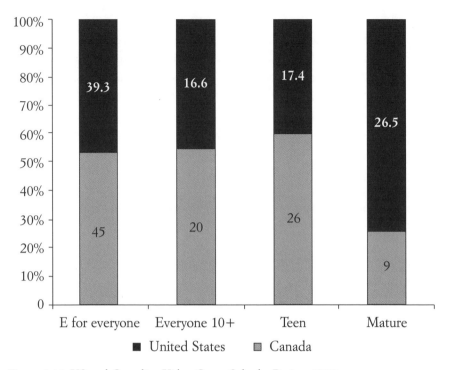

Figure 2.12 US and Canadian Video Game Sales by Rating, 2012
Note: US data incomplete for 'Mature' sales.
Sources: ESA (2012); ESA Canada (2012).

It's notable that in the US example, video game ratings tend to focus on vio-lence, though an increasing number of cases are beginning to regulate sexual content. In most other countries, however, games are not only more strictly reg-ulated in terms of a broader range of content, but that regulation typically entails government involvement. In fact, in the last six years, a number of countries, including Germany, Japan and Australia, have overhauled their ratings systems. In the US, demands for stricter controls in light of recent game scandals, such as the discovery of sexual content hidden in the code of *Grand Theft Auto: San Andreas*, the industry has even incorporated a new rating (Wong, 2005). Ironically, sex has a long history as part of the content and marketing of video games, *BMX XXX* (2002) having been the first major title to include live-action nudity (Tran, 2002). However, in 2004, games rated 'Mature' ('M') increased dramatically, by 12 per cent from the previous year alone (Wingfield and Marr, 2005). Like the MPAA, the ratings system includes Canada in the American mar-ket, and puts most of the onus for content on the consumer. It's also worth noting that the content descriptors used by the ESA have no direct relationship to the ratings themselves; they are purely for the consumer's reference.

Agency	Country/ region	Description
Australian Classification Board	Australia	State two-tiered system of 'advisory' and 'restricted' ratings. 'Advisory' ratings consist of four colour-coded age-appropriateness ratings, while there is currently only one colour-coded 'restricted' rating.
Computer Entertainment Ratings Organisation (CERO)	Japan	State system of five age ratings descriptors combined with seven mutually exclusive content descriptors. In addition two special categories exist for statistical software and trial versions.
Entertainment Software Ratings Association (ESRA)	Iran	State system of six colour-coded age-appropriateness ratings.
Entertainment Software Ratings Board (ESRB)	US and Canada	Industry system of six age-appropriateness ratings combined with thirty overlapping content descriptors.
Game Ratings Board (GRB)	South Korea	State system of four colour-coded age-appropriateness ratings combined with seven mutually exclusive content descriptors.
Office of Film and Literature Classification (OFLC)	New Zealand	State system of three colour-coded tiers of ratings – restricted (Yellow), unrestricted (Green) or banned (Red) – combined with recommended age levels.
Pan European Games Information (PEGI)	European Union	Voluntary state system of five age ratings classifiers and eight mutually exclusive content classifications. EU countries can opt in or use their own ratings system.
Unterhaltungs-software Selbstkontrolle (USK)	Germany	State system of five colour-coded age-appropriateness ratings.

Table 2.9 State Institutions Involved in Game Ratings, 2012

Sources: Australian Classification Board (2008, 2010); CERO (2008); ESRA (2010); OFLC (2008); PEGI (2007); USK (2007, 2010).

Moreover, the Interactive Entertainment Merchants Association (IEMA), an organisation representing 85 per cent of game retailers, requires proof of age from anyone purchasing games rated 'M' (Thorsen, 2003). According to the ESA and some parents' groups, the system is working reasonably well; in 2004, 46 per cent of all games sold were rated 'E', suitable for everyone. As noted earlier, there is also a distinction between platforms, with 64 per cent of all games for Nintendo Gamecube carrying an 'E' rating (Felberbaum, 2005b).

Ironically, most attempts to pass state laws in the US to ban the sale or rental of violent video games have been struck down as violating free speech rights (Tallon, 2005). Other countries, however, have not had the same difficulties. It is also worth considering that the game ratings systems are voluntary in some places. For example, in the European Union, not all countries subscribe to the EU system, but instead follow their own or none at all. Perhaps most interesting is the response from the video game industry, particularly the US segment. In most cases, government systems are viewed as complex, though many offer fewer content categories than the US system as well as categories that are mutually exclusive. Government involvement in game ratings in other countries has allowed for stricter control of content. In New Zealand, for example, the Office of Film and Literature Classification banned its first video game *Manhunt* ('Computer Game Banned for Repetitive Violence', 2003). Little information is available on the industry's efforts to self-police outside the US and Canada, however. More significantly, however, the UK government has taken the lead in actively supporting its video game industry though other governments have made similar moves; such government support – typically through subsidies and tax benefits – has the potential to make subsidised regions, companies or types of workers more competitive in the global market (Spectrum Strategy Consultants, 2001). Similar moves have been discussed in Australia and New Zealand.

Most ratings systems are based on similar ones used in film and other media, highlighting one more way in which the industry owes its development to the logics of film. What is important to note, however, is that in many cases – particularly in the EU and Japan – ratings descriptors go well beyond simple age recommendations, giving clear indications of which types of content one might find in particular games. Table 2.9 details the agencies responsible for the ratings systems in Australia, the European Union, Germany, Japan, Korea, Iran New Zealand, the US and Canada. In most of the systems outside North America, the systems are state-run, though it should also be noted that some are voluntary. In most cases, there is a combination of coding systems including age-appropriateness, colour-coding and mutually exclusive content descriptors. The response from industry officials, particularly in the US, to these more complicated systems has been mixed. Typical rhetoric describes them as overly complex, though the additional coding could be useful to customers assessing age guidelines and content suitability.

Beyond regionally specific regulation concerns, video games face many of the same problems as other products. Piracy has begun to plague the industry almost to the same degree as in the film business. Software pirates from Russia, now one of the global leaders in piracy, have cost US businesses an estimated US $1.7 billion in losses (Chazan, 2005). There is also the problem of online file sharing, with services like BitTorrent taking parts of files from various computers connected to the network. Interestingly, though BitTorrent shares not only game but also movie and music files, it has only received a single legal complaint so far (Veiga, 2004).

The question of intellectual property has cut both ways in the video game industry. Because many games, particularly MMORPGs, result in players designing and creating content while they play, the question of ownership has been muddled. While there has been little legal consequence in the US, with gamers mostly selling their 'creations' through sites such as eBay, in other countries, legal action has resulted (Castronova, 2001, 2002). In China, a software company was forced by the courts to compensate a player for lost 'property' on its game. The player had spent more than 10,000 yuan and two years with the game when his creations were stolen by a hacker who exploited loopholes in the code (CNN, 2003). Should similar action be taken elsewhere, it could have serious ramifications for an industry that has come to view online games as a great potential source of revenue. It also would suggest a shift in what is meant by labour, an idea taken up in Chapter 5.

Finally, a number of recent industry-specific legal concerns have impacted video games. A number of companies, but most notably Electronic Arts, have been hit with lawsuits from employees citing unfair labour practices (*Jamie Kirschenbaum, Mark West, and Eric Kearns v. Electronic Arts, Inc.*, 2004). These cases will be given more attention in Chapter 5, but are significant for their potential impact on regulation. A second concern, not uncommon to the computer and software industries, has been the question of patent violations. The most recent of these charges has centred around Sony, which has been seeking damages from companies in Australia that have provided hardware that circumvents Sony's copy-proofing technology. Sony was also forced to pay US $90.7 million in damages after it was found that its PlayStation controllers violated another firm's patent (Fordahl, 2005). Similar infringement suits have been filed over software. Sega sought damages from Fox Interactive, citing similarities between the latter's *The Simpsons: Road Rage* (2001) and Sony's *Crazy Taxi* (1999) (Fahey, 2003).

Because so much public concern about video games has centred on violent content, it seems likely that the focus of regulation – both within and outside the industry – will continue to deal with these factors rather than with production and struggles over labour and patents. Because there is considerable overlap between the industry's concerns and those of other media industries, a more detailed analysis of how video games are tied to these industries is in order.

CONCLUSION

In the modern era, the video game industry has been successful in courting and cultivating a wide audience for its products, while establishing a stable industrial structure that has, thus far, avoided the problems that plagued the business in the 1970s and 1980s. Since the 1990s, the industry divided into four primary sectors: software development, software publishing, hardware manufacture and retail. The stability achieved has largely been the result of the development of software publishers, who serve as gatekeepers and quality control between hardware manufacturers and small game development studios. Hardware manufacture and retail are taken up in the next chapter.

The power given to hardware publishers, however, has led to high levels of concentration, with a small number of publishers making up a majority of the industry's profits globally. This, in turn, has allowed them to pick and choose developers to establish formal ties with, often through direct acquisition of those companies. It has also given them tremendous power in directing the industry, and it is no coincidence that all four of the most profitable concerns in the industry are directly involved in software publishing.

At the same time, the industry has begun to actively try to expand its audiences, by experimenting with new formats like online gaming, casual games and wireless gaming. As Chapter 4 will show, the industry's use of licensing has been another key attempt to reach larger audiences. These experiments have not been without cost, however. While casual gaming and wireless gaming have become increasingly profitable, they have also expanded the ways in which games can be distributed. This development is something that console makers in particular have been forced to adapt to. This will be discussed in more detail in Chapter 3.

One other significant development has been the growth in the number and prominence of international game developers. In part, this is because of the popularity of video game products, but it is also due to the predominance of international companies in electronics manufacture, computer and microchip manufacture. Many of these companies will be discussed in more detail in Chapter 3. This internationalisation has gone hand in hand with the expansion of video game audiences, but it has also resulted in an increasingly complex system of regulations which game developers must struggle with. Because most of this regulation has involved government ratings systems, the industry has worked to try and advance its own view of how games should best be regulated.

Finally, the video game industry's development has a number of parallels with the film and recorded music industries. Having secured a distinct niche for its products in the cultural marketplace, it has begun to try to expand its purview, working to form ties with other media industries, both through cross-ownership and through licensing and partnerships. These matters will be taken up in Chapters 3 and 4.

3

Video Game Hardware, Distribution and Retail

While software has been the most profitable sector in the video game industry, the importance of hardware cannot be underestimated. The term hardware refers to any of the tools needed to play video games. These tools might include consoles, handheld games, arcade games or even personal computers. In addition, hardware includes peripheral equipment such as joysticks, memory cards, network adapters, headsets, keyboards, etc. In the early history of video games, hardware and software came as a combined package. But, with the exception of arcade games, as technology progressed, software and hardware separated.

As Chapter 1 demonstrated, hardware development has proved key to the industry's advancement because it has pushed the technical capabilities of games from the earliest days, while in the modern industry, hardware has served as a loss-leader to help defray the high costs of new software purchases. It is no coincidence that three of the major software publishers are also hardware producers: Microsoft, Nintendo and Sony. Controlling the platforms on which games are played has been a key strategy, and the matter of distribution is becoming increasingly central to the industry, necessitating close examination of retail and of digital distribution methods in addition to an understanding of the platforms themselves.

While Moore's Law about the pace of microchip development has been a crucial factor in the hardware sector, driving the planned obsolescence that allows both for software advances and continued consumption of new products, matters of convergence involving a single piece of media technology taking on the function and capabilities of others have also proved important (Pavlik, 1998). Microchip manufacture, too, has come to be dominated by a small number of companies. Among these are IBM, nVidia, Intel and AMD. As seen in Chapter 1, arcade games and the earliest video game consoles were single function machines but, as hardware manufacture has advanced, the machines have evolved into entertainment hubs, allowing players to use a variety of media forms, from CDs and DVDs to the Internet and telecommunications. Similarly, personal computers are being designed with increased cross-platform capabilities, including the ability to record TV programmes, display images and manage household functions (Gnatek, 2005).

At the same time, it is in the hardware sector that new ways of bringing gamers together is being tested. Among these ideas are online gaming via networked computers and various new controller technologies. Video game and technology giant Sony was one of the first to experiment with networked gaming, using mainframes with partners Butterfly.net and IBM, to test the capabilities of MMORPGs (Bulkeley, 2003). While such an experiment was reminiscent of the early days of the Internet and the Advanced Research Projects Agency's network, it was significant because it devoted mainframes, some of the most expensive computer hardware available, solely to video games. Estimates projected that such a network could allow as many as 1 million players to join in a game simultaneously. The network capabilities required to run modern MMORPGs are immense. The hit game *World of Warcraft* runs on a system of global servers, with more than 75,000 central processing unit (CPU) cores and more than 112.5 terabytes of memory (Gamasutra, 2009).

Another example of the significance of the video game hardware sector can be seen in its impact on other media technologies. Sony's PlayStation 3 features both a new form of computer processor and the use of Blu-ray optical discs (Guth and Divorack, 2005). The Blu-ray disc was one of two formats touted as the next step up from DVDs; the battle over formats resulted in a number of strategic alliances in Hollywood and with major retailers. Ultimately, Blu-ray won out, though the technology's adoption has been slow (Kiss, 2008). In spite of the inclusion of Blu-ray technology in the PlayStation 3, or perhaps because of the cost of the console due to its inclusion, PlayStation 3 sales have lagged behind those of Sony's previous console, the PlayStation 2 (Gamasutra, 2008).

The reach of consoles is changing as well, adapting to roles other than gameplay. The National Center for Supercomputing at the University of Illinois at Urbana-Champaign created a supercomputer composed entirely of PlayStation 2s. Researchers found such a computer particularly adept at rendering digital displays, and so potentially valuable to institutions like the US Defense Department, where graphical representation is particularly important (Markoff, 2003). Markoff notes that there has been a shift in recent years, with the fastest computer capability being developed not directly for government or business, but rather in response to consumer demand. This marks a destabilisation in the chip market which, as noted in Chapter 1, had been a historical problem for the video game industry. It also represents a growth for the semiconductor industry, which saw chip sales increase by almost 3 per cent in 2002, in part due to the increased demand for gaming systems (New York Times, 2002).

It is in the hardware sector that the second crucial example of the transnational nature of video game production is seen. While most game hardware and software is sold in North America, Western Europe and Japan, a significant proportion of hardware is contracted out to nations where game consoles are considerably less affordable. This fact has led to competition to craft a console recognising the needs of the remainder of the global market (Aslinger, 2010). The production of memory and microchips in particular, is contracted out to nations including China, India and Taiwan (Lüthue, 2006). Further, many of the mineral resources required come from struggling nations, including the Congo (Game Politics.com, 2008; Lasker, 2008; Plunkett, 2010). Typical analysis of the microchip industry has looked at their manufacture only in terms of mainframes and personal computers, where high-end chips are crucial to the effectiveness of video game software. However, as consoles have become more popular, their impact on the microchip industry has grown. Typically, microchip manufacture is contracted out from console makers to outside firms, making cost estimates difficult. In fact, as the industry has grown, outsourcing across all sectors from hardware to software has become increasingly common. This has resulted in a number of benefits. First, it reduces the cost of labour while making round-the-clock production possible. Second, it has allowed the industry to seek out less stringent environmental regulation (Kerr, 2006a). Microchip production involves highly toxic procedures, whose impacts on workers can be severe (Byster and Smith, 2006). Moreover, contracting out production makes unionisation much more unlikely (Steiert, 2006). The difficulty of unionisation will be discussed further in Chapter 5.

What all of these examples prove is that, while the hardware sector might not be the most dynamic part of the industry, it has more impact than it is often credited with. It is hardware that provides the most effective limits on gameplay, regardless of the style of game. While there have been developments across all areas of the hardware sector, the most significant changes are happening within consoles. In particular, the adoption of consoles as sites of media convergence has resulted in some pushback from other media industries, particularly telecommunications, where wireless smart phones have increasingly functioned as game consoles in their own right.

This chapter focuses on the hardware sector in order to highlight the importance of consoles in the gaming experience and in maintaining the situation whereby the industry is controlled by a small number of companies, two of which are detailed in case studies. It also examines the growth in handheld and mobile gaming, in order to contextualise the discussion of digital distribution. It concludes by looking at the role of arcades and the retail sector's ties to the overall industry.

THE HARDWARE SECTOR

From the mid-1990s, the console sector has been both highly profitable and highly concentrated. In 2000, Sega pulled out of the sector, and since then the console business has been dominated by three companies: Microsoft, Nintendo and Sony. Over the following decade, all three jockeyed for the lead position. In 2004, more than 120 million consoles of various sorts were sold worldwide (Wingfield, 2005). By the end of 2008, that number had increased to more than 246 million consoles in total (VGChartz, 2008b). Figure 3.1 displays changes in global market share of the major companies' console sales from 2001 to 2010. It includes any major platform, including consoles and handhelds. The numbers are impressive not just in their own right but because they are starting to rival other technologies such as DVD players. This is due, in part, to the fact that consoles can now serve other functions beyond gameplay. While this shows con-siderable volatility within the industry, the same companies have dominated and pushed others out.

Typically, the changes in market share have been tied to the number of plat-forms offered by a company at a particular time as well as the timing of their release of a new console. In 2001, Microsoft had just released its first console, the Xbox, and by 2005 had both the original and the newly released Xbox 360. By 2006, both Nintendo and Sony had answered with their latest, respectively,

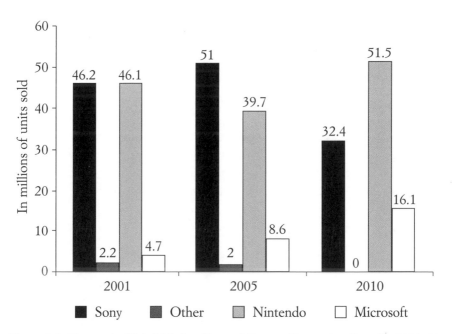

Figure 3.1 Changes in Global Market Share of Current Generation Consoles 2001–10
Source: VGChartz (2010a).

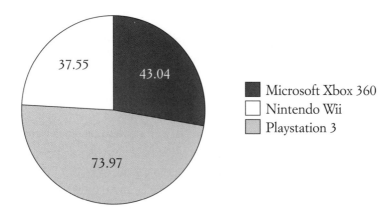

Figure 3.2 Total Worldwide Sales of Seventh-generation Consoles, in Millions of
         Units, 2010
Source: VGChartz (2010c).

the Wii and the PlayStation 3. Figure 3.2 shows the market share of the most
recent console generation: Microsoft's Xbox 360, Nintendo's Wii and Sony's
PlayStation 3. Releasing the Xbox 360 into the marketplace before the other
new consoles enabled Microsoft to establish a strong market share, which has
since been chipped away, primarily by Nintendo's Wii. All three have considerable resources and have developed highly loyal fan bases, but some analysts have
noted that it may be difficult for the market to support more than two consoles
over any lengthy period (Pereira, 2002b). The question of how many consoles can
be supported has been made trickier by continued demand for consoles from
the previous generation. Table 3.1 details the top-selling consoles by total units
sold worldwide, demonstrating the popularity of some earlier consoles and relatively weak sales of some of the most recent releases. The Wii, for example, has
sold tremendously well. It also points to some of the difficulties of maintaining
multiple consoles on the market. Sony, in particular, has seen continued demand
for its PlayStation 2 console, which has still sold the most worldwide. By
September 2010, it had sold more than 137 million units worldwide, more than
any console or handheld (VGChartz, 2010c). Sony's strategy to keep two
platforms on the market – the PlayStation 2 and the more recent PlayStation 3
– has been questioned, particularly as the PlayStation 3 has lagged behind
(Watson, 2008).

    High demand for consoles is often a driver for sales, as well. In 2003, Sony
and Microsoft dropped their prices in order to help boost sales. The tactic
worked, and by 2004, increased demand combined with difficulty meeting it
enabled Sony and Microsoft to keep prices of the PlayStation 2 and the Xbox
higher (Pereira, 2002b). While cost cuts had an impact, high demand and higher

Platform name and year of release	Type of platform	Units sold
Sony PlayStation 2 (2000)	Console	137.05
Nintendo DS	Handheld	134.06
Nintendo Gameboy	Handheld	118.69
Sony PlayStation (1994)	Console	102.49
Nintendo Gameboy Advance	Handheld	81.49
Nintendo Wii	Console	73.97
Nintendo Entertainment System (NES)	Console	61.91
Sony PlayStation Portable (2004)	Handheld	59.91
Super Nintendo Entertainment System (SNES)	Console	49.10
Microsoft Xbox 360 (2005)	Console	43.04
Sony PlayStation 3 (2006)	Console	37.55
Nintendo 64	Console	32.92
Sega Genesis	Console	28.54
Microsoft Xbox (2002)	Console	24.65
Nintendo Gamecube	Console	21.75

Table 3.1 Top Fifteen Hardware Platforms Worldwide by Millions of Units Sold
Note: sales estimated through 13 July 2010.
Sources: Market Share Reporter (2010b, 2010c, 2010d, 2010e, 2010f, 2010g, 2010h, 2010i); VGChartz (2010a, 2010c).

prices weren't the only way console makers offset such losses. Console makers typically earn between $5 and $10 in royalties per game sold on their consoles (Grover *et al.*, 2005). As discussed in Chapter 2, console makers license publishers and developers to create games for their consoles, and part of the licensing agreement earns them a percentage of video game sales. The more games available for a console, the higher the royalties they stand to make. This system has certainly been beneficial, but it hasn't guaranteed the companies' success. Microsoft took heavy losses related to the original Xbox, which necessitated tremendous changes in the production model for the Xbox 360. Some estimates suggest Microsoft lost as much as $1.2 billion per year on the Xbox (Guth *et al.*, 2005). In contrast, Nintendo struggled because it had a younger audience, particularly as the demand for game realism increased and its original consumers aged (Pereira, 2002b). Sony's problems have been twofold. Internally, the company has struggled with shaky finances and with leadership turnover (Guth *et al.*, 2005). In addition, as previously noted, Sony's PlayStation 3 was the last of the most recent console generation to enter the market, and it was saddled with a high cost because it incorporated a number of additional technologies.

The push towards convergence is one of the most notable features of the current lineup of consoles. By including Internet connections allowing consumers to access downloadable content, including older games, the systems have a sort

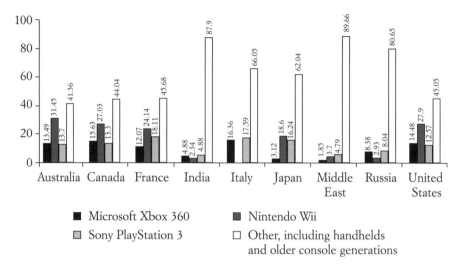

Figure 3.3 Comparison of Console Market Share by National and Regional Markets, 2009

Sources: Market Share Reporter (2010b, 2010c, 2010d, 2010e, 2010f, 2010g, 2010h, 2010i); VGChartz (2010a).

of backwards compatibility, limited to the console maker's willingness to provide games from previous consoles. This would seem to give Sony the advantage, as its PlayStation 2 has more than 2,000 games designed for it in addition to those made for the more advanced PlayStation 3 (Grover *et al.*, 2005). They can also serve as media hubs. All three consoles have the ability to play movies streamed from the Internet via services such as Netflix. In addition, both the Xbox 360 and the Sony PlayStation 3 are HD compatible and can play DVDs and CDs, as well as store data on their hard drives (Taub, 2005). They can also download television shows and communicate with other networked consoles, allowing players to talk as they play (Tran, 2002). For both Microsoft and Sony, the inclusion of a hard drive drove the costs of the console up. Sony added to the burden by including Blu-ray technology. Sony also developed an entirely new processor, called the Cell processor, with partners Toshiba and IBM; these processors are reportedly ten times as powerful as most standard computer processors (Slagle, 2005c). In spite of this, Nintendo's Wii, which has no HD or DVD capability, has been the bestseller of all three consoles since its introduction (Peckham, 2008; VGChartz, 2008a, 2008b). Figure 3.3 shows the market share in 2009 of major consoles in a number of international markets. While the data cover a small number of countries, they clearly indicate both a pattern of adoption that is highest in North America, Western Europe and Japan. Earlier consoles and handhelds are still very popular elsewhere and make up a significant portion of the market, but the North American, European and Japanese markets accounted for approximately 80 per cent of global video game hardware consumption (Nichols, 2013). Adoption of the latest consoles outside those regions has proved very slow. It also indicates some national preferences that have been suggested elsewhere, particularly in the case of Japan, where Microsoft's video game products have struggled.

The competition for market share has been fierce. In 2002, both Sony and Microsoft entered into an intense price war that established practices for most of the decade. In the US market, both companies originally priced their consoles at US $299, but then entered into a price war that resulted in a price drop on both to US $179 (Musgrove, 2001; New York Times, 2002). Essentially, consoles became loss-leaders, forcing the industry to make most of its profits from software sales. Microsoft also spent more than $200 million marketing the Xbox console in 2001 to help push demand for its release (Pereira, 2002b). In spite of this, Microsoft lingered in second place in all markets, something it had to work to address with the release of the Xbox 360, though it remained stuck in second place globally and in most national markets (Peckham, 2008). In addition, it's worth noting that San Diego-based Qualcomm and Tectoy, a Brazilian firm, have released an alternate console – the Zeebo. This console began shipping during 2009 in Brazil and Mexico, and was launched in India in 2010, with plans to launch in Russia and some Asian markets as well, but by the end of 2011, the company had all but ceased production and was retooling its operations (Wauters, 2011). The venture was not designed as direct competition to the major consoles but to address a market the seventh-generation consoles were neither well designed nor priced for (Aslinger, 2010).

Analysts have been uncertain how best to explain the video game market. Prior to the release of the current generation of consoles, many assumed that Sony would continue its dominance. Others suggested that being the first to launch a new console, as Microsoft did with the Xbox 360, would be enough to propel a console into market dominance. One estimate suggested that getting the Xbox 360 into the marketplace before Sony or Nintendo could release consoles would give Microsoft a 38 per cent market share while Sony's would drop to 32 per cent (Slagle, 2005b). However, it was the Nintendo Wii, with its minimal nods to convergence, which bested allcomers, and proved to be a product retailers were unable to keep on the shelves.

CASE STUDY: MICROSOFT

Of all three firms, Microsoft had perhaps the most to overcome. When the company moved from production of the Xbox to the Xbox 360, it tried to promote convergence in terms of both its hardware and the software licensed for its machines. Known for its operational and productivity software development, Microsoft has had less success with its attempts at video games. The company both manufactures platforms and publishes video games in addition to its other ventures. Its ten-person board of directors has ties to J.P. Morgan, AT&T, BMW and a number of investment banks. The company was founded in 1975 and has had extensive dealings in video games, manufacturing software long before entering the console sector (Hoover's, 2005; Microsoft, 2005). Table 3.2 details its major software and hardware franchises and their years of release.

Microsoft may often be seen as the most recent company to enter the video game hardware industry but it actually has a much longer history in the product than many believe. This foray into games consoles is one of several ventures designed to leverage Microsoft into the media content business, with the hope that its consoles, which also allow access to other media, will function as enter-tainment hubs. The Xbox platform was launched in 1999 and the company has struggled to make a profit from its video game ventures in spite of the popularity of the *Halo* franchise. Though well received by players, the console experienced some problems. Many of these related to the supply chain, much of which was out of Microsoft's control. For example, its use of 'off the shelf' Intel chips left it subject to heavily fluctuating market prices (Daily Gleaner, 2005; Guth, 2005). The cost of components is believed to have lost the com-pany $125 for each console sold (Takahashi, 2005). Miscalculations about the cost of chips and hard drives cost Microsoft $4 billion between 2001 and 2005 (Guth, 2005).

These losses led to a change of tactics when it came time to release the next console. According to Todd Holmdahl, vice president of Microsoft's Interactive Entertainment Business, they quickly learned that 'you want to control your des-tiny'. This meant 'controlling the price, controlling the margins, and controlling the designs' (Takahashi, 2005). To do this, the company started by changing its software development strategy, employing more than 900 programmers dedi-cated to software design for the Microsoft console and PC game offerings. The expectation was that the combination of new consoles with better software would drive demand (Marriott, 2005a). This change in console capability meant a rise in prices for both software development and to consumers, one of the chief difficulties video games face (Wingfield, 2005). Such costs were risky, par-ticularly for an industry still battling with the perception that its products are meant for children (Gentile, 2005b).

Released in 2005, the most recent console version, the Xbox 360, was envi-sioned as a home entertainment hub. The use of the Windows operating system was one of the key changes, though it is difficult to determine whether this helped or hindered the console's fortunes. One of the first steps to overcoming

Year formed	**1975**
Headquarters	**Redmond, Washington, US**
2012 Sales (US$m)	**73,723.0**
2012 Employees	**94,000**
Industry sector	**Hardware, software publisher**
Hardware and year introduced	
Consoles	Xbox – 2002
	Xbox 360 – 2005
Non-game hardware	Zune (digital media player) – 2006
	UltimateTV (DVR box, discontinued) – 2000
Key software franchises/games	
Game software	*Halo*
	Flight Simulator
	Black and White
	Perfect Dark
	Age of Empires
	Age of Mythology
Non-game software	*Windows*
	Windows 95
	Windows XP/2000
	Windows Vista
	Microsoft Office
	Microsoft Development Tools
	Windows Mobile
Studios owned	ACES Game Studio – Redmond, Washington, US
	Carbondated Games – Redmond, Washington, US, Hyderabad, India, and Beijing, China
	Ensemble – Dallas, Texas, US
	Hired Gun – Redmond, Washington, US
	Lionhead – Guildford, UK
	Rare – Twycross, UK
	Turn 10 – Redmond, Washington, US
	Wingnut Interactive – Wellington, New Zealand
	Massive, Inc. – New York, US

Table 3.2 Corporate Profile of Microsoft Corporation, 2012
Sources: Hoover's (2012); Mergent (2008).

the challenges faced by the first console was to distinguish the console and its games in the market in terms of realistic graphics and sound. This was achieved by making games and the new system HD compatible. Microsoft estimates suggested that by 2008, more than 100 million homes in the US would have HD compatible TV sets, and the company hoped to take advantage of this fact

(Snider and Kent, 2005). While the number was considerably lower – somewhere between one-quarter and one-third of American households had HD televisions in 2008 – the realism achieved helped to distinguish the Xbox 360's games from the next console to enter the marketplace, the Wii (Albrecht, 2008; Nielsen, 2008a). By 2010, however, HD TVs were estimated to have been adopted by approximately two-thirds of American households (Clabaugh, 2010). In addition, the Xbox 360 console is able to stream videos and pictures from Windows compatible devices and digital cameras. Like Sony's PlayStation 3, the Xbox 360 also features a processor capable of making it more powerful than many home computers (Marriott, 2005a). In keeping with the desire to create an entertainment hub compatible with Windows computers, even the console's navigation system is designed to mimic that of Windows (Bajak, 2005).

The most significant changes take place before the consoles reach the audience. Supply chain issues, particularly in relation to the console's microchips and hard drives, made the original Xbox a drain on company coffers. When it was time to produce the Xbox 360, there was a change of strategy. First, Microsoft attempted to work with both its close partner Intel and nVidia on a chip design unique to the Xbox 360 that Microsoft could own, but neither firm was willing to allow Microsoft ownership (Takahashi, 2005). Instead, IBM signed a deal with Microsoft to create a unique chip for the company as it had done for both Nintendo and Sony (Bergstein and AP, 2006a). By gaining control of the chip, Microsoft was able to avoid fluctuating chip prices. Similarly, the company changed its use of hard drives, including a detachable hard drive that cheapened the cost of the consoles themselves even as it allowed for multiple sales opportunities down the line (Guth, 2005).

Second, the company centralised its production of the consoles themselves. It opened two factories in southern China run by competing manufacturers as a safeguard against supply chain problems. In addition, many of the other parts that make up the console are manufactured near the assembly plants, thus minimising costs and bypassing Chinese import rules and tariffs (Guth, 2005). For Microsoft, finding ways to take advantage of globalisation is key. According to Larry Young, 'Globalization's a big part of this … [it's about] leveraging partners' (Takahashi, 2005). This is impressive because the console is made up of more than 1,700 individual parts, most of which are produced in fairly close proximity to the assembly factories in China. These parts come from more than 250 suppliers using more than 25,000 employees (Guth, 2005).

Third, the company focused on building better relations with game developers, particularly in Japan. At the time of the console's release in 2005, forty-five titles were slated for release in Japan and thousands of game design kits had been dispatched to developers, while estimates suggested rival Sony had only sent out hundreds (Takahashi, 2005). Finally, the company tried to leverage the rise of HD televisions and the console's early release into marketing opportunities, as in a deal with electronics manufacturer Samsung for store demonstrations of the consoles in conjunction with Samsung's newest HD televisions (Economist, 2005).

HANDHELDS AND MOBILE GAMING

The second major platform for video games is the handheld. Like consoles, handhelds have a long and successful history. Unlike consoles, however, the business has been dominated until very recently by a single player: Nintendo. Until almost 2005, Nintendo controlled almost 98 per cent of the global hand-held market (Slagle, 2005a). Nintendo released the Gameboy, its first handheld, in 1989 and, as discussed in Chapter 2, it is one of the major innovations that revitalised the industry. More recently, the company released the Gameboy DS in 2004, selling more than 1 million units by the end of the year, a majority of these sales during the holiday season. This represented a rapid adoption even by current standards; for example, it took Apple's iPod nineteen months to reach the million-unit level (Biersdorfer, 2004). Like consoles, handhelds have embraced convergence technologies, and have the ability to play music, television and movies (Tran, 2002). Users can also connect and play games with up to fifteen other players in their vicinity through the use of Wi-Fi transmission (Felberbaum, 2005).

In 2005, Sony entered the handheld market, hoping to cut into Nintendo's market share. With its PlayStation Portable (PSP), it attempted to draw a wider audience for handhelds. Nintendo's units had been targeted primarily at children but Sony hoped the PSP would bring handhelds to a more adult audience. Since the PSP's introduction, the competition in the handheld market has grown fierce, and Nintendo's market share dropped to 67 per cent by 2008 (VGChartz, 2008b). Since this time, market share has remained fairly constant globally but varies greatly from country to country. Figure 3.4 compares the penetration of handheld devices by country. Clearly, the handheld market is highly concentrated, the majority of all sales happening in the US and Japan. There are, however, a considerable number of handhelds sold elsewhere. When sales of specific types of handhelds are viewed (Figure 3.5), Nintendo is still clearly the leader, though the PSP has made in-roads, particularly in European countries.

Handheld and mobile gaming represents an interesting crossroads for the video game industry. Perhaps owing to hard economic times, the games market began to slow in 2008 (Business Wire, 2009). In 2009, console games dropped from 79 per cent to 71 per cent of global game sales, while mobile sales grew. The development of smart phones has marked a major change,

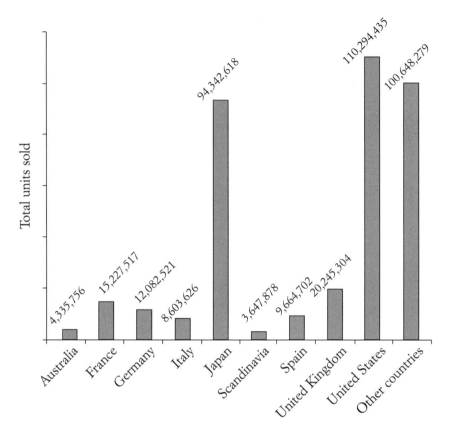

Figure 3.4 Total Handheld Units Sold by Country, 1989–2010
Source: VGChartz (2010b).

introducing new platforms and distributions systems. In fact, iPhone alone
accounted for 5 per cent of video game sales in 2009 (Stone, 2010). One survey
even suggested that the mobile phone has become the preferred gaming platform
for adults in the US and UK, with 52 per cent of respondents in the UK indicating that they had played a game on their devices, against 44 per cent in the
US (Baar, 2011). However, the casual game market, which has become increasingly tied to the handheld and mobile gaming hardware sector, has been forecast
to grow to at least US $11.7 billion by 2010 (Computer Weekly News, 2009). A
significant portion of this growth, however, uses hardware outside the video game
industry's control: cell phones, particularly smart phones like Apple's iPhone,
Nokia's nGage and Google's Android (Entertainment Newsweekly, 2009). While
it is expected that portable game systems designed by the major manufacturers
will continue to dominate, their growth seems to have peaked. Apple's iPod and
iPhone sales are estimated to represent almost a quarter of all portable game

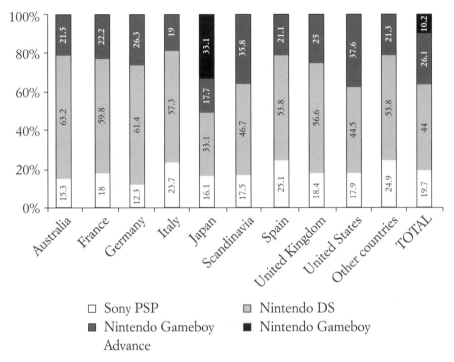

Figure 3.5 Comparison of Handheld Market Share by Country, 2010
Source: VGChartz (2010b).

sales currently, something which the major handheld manufacturers are taking note of (Computer Weekly News, 2009). Apple's decision to lower the price of the iPod Touch in 2009 was seen as a move to position it as a viable gaming platform, an area Apple has struggled with in relation to its personal computers (National Post, 2009).

Part of the reason for the surge in video games on cell phones is due to the difficulty in negotiating deals with telecom corporations. Both Sony and Microsoft have previously attempted to work with cell phone manufacturers but with little success. The current wave of smart phone development, however, has allowed game makers to bypass the telecoms in favour of working with distribution systems like the Apple and Google application stores. This strategy has made the creation and distribution of mobile content considerably less onerous (Economist, 2009). Mobile platforms also make life easier on developers. Many emerging game developers are focusing on mobile games, emphasising affordability and ease of play over the more traditional graphic realism and franchise content (Wolverton, 2010). But studies show that smart phones and tablets are both poised for rapid growth in the coming years. Figure 3.6 shows the projected growth of both through 2015.

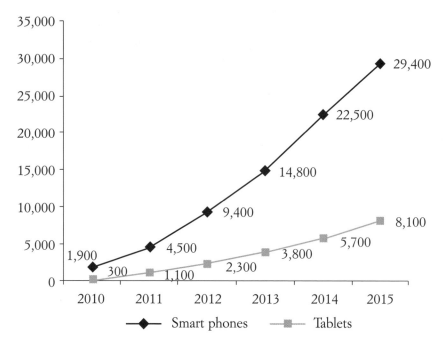

Figure 3.6 Global Smart Phone and Tablet Shipments 2010–15 (US$m)
Source: Bilton (2011).

Apple's success has been seen as the primary reason for the explosion in mobile games, because of the tremendous sales of the iPhone, the iPod Touch and the 2010 release of the iPad (Stone, 2010). Games on the iPhone are more popular than on most other smart phones, with estimates suggesting that they're four times as popular on the iPhone than the Android and twice as popular as the Blackberry (Business Wire, 2009; PR Newswire, 2010a). The popularity of the iPhone and iPod Touch is impressive. By September 2009 more than 30 million iPhones and 20 million iPod Touches had been sold worldwide, and by the following year the two devices had outsold the Nintendo Wii globally, giving Apple the potential to be a serious contender in the video game market (National Post, 2009; Wolverton, 2010). By June 2009, Apple's distribution service, the App Store, had sold more than 50,000 apps and had had more than 1 billion downloads (Tedesco, 2009). Game sales accounted for US $500 million in revenue for Apple in 2009 (Bulik, 2010). Of all the apps available, games are the most downloaded of any categories (Kim, 2010). In 2010 game apps represented not just the number one category of downloads but as much as 17 per cent of all apps downloaded (Bulik, 2010).

The rate of growth for iPhone users is impressive. One study found that 15 per cent of consumers in North America and Europe owned either an iPod

Touch or an iPhone by 2009, while 29 per cent owned a Nintendo DS (Business Wire, 2009; Computer Weekly News, 2009). More tellingly, at that time almost 22,000 games were available for the iPhone and iPod Touch, against only 3,700 for the DS and fewer than 1,000 for Sony's PSP (National Post, 2009). In the same survey, more than half of North American consumers and almost 70 per cent of European consumers indicated that they had played a game on their mobile phone. Both groups showed high willingness to purchase games as well, with 45 per cent of North American consumers and 36 per cent of European consumers having paid for mobile phone apps, including games (Business Wire, 2009; Computer Weekly News, 2009). Convincing consumers of the worthiness of these devices as gaming platforms may be difficult, however. When the iPad was released, only 30 per cent of users said they would use it for gaming (Bulik, 2010). But the iPad represents a second potential for game users. Some have begun to suggest that tablets and smart phones could serve as game controllers if enabled to connect to TVs (Kim, 2010). Such a move suggests interesting possibilities for game design, and something that the major industry players must consider. As one game developer put it, 'The tablet shifts [the casual game] battle to the living room', an area that has been the traditional domain of the major console manufacturers (Kim, 2010).

The main software developers are paying increasing attention to mobile games. Some, like EA and Ubisoft, have begun to develop apps in hopes of using them as a gateway to console and PC game sales rather than as a major driver of revenue in themselves (National Post, 2009). In part, this owes to the cost of apps themselves. Users of mobile apps have become accustomed to a two-tiered system, in which many apps are free and the rest tend to cost between US $0.99 and $5 (Economist, 2009). Often free apps are versions of paid apps with reduced functionality, designed with a loss-leader intentionality, hoping to convert users to the more advanced application. Conversion, however, is rare, ranging from between 2 and 13 per cent on average. While this may seem low, it suggests that members of the eighteen–thirty-four-year-old demographic might be valued at approximately US $25 per 1,000 users of a mobile game or application, a number roughly equivalent to their value to network television (Economist, 2009). In spite of the lower price point, some major developers have had success with higher-cost mobile games. In 2009, for example, both Sega and EA experimented with higher-cost games. Sega released an iPhone version of *Sonic the Hedgehog* in May 2009 and in June EA released its version of *Sims 3* with a US $9.99 price, which still managed to shoot to the top of the app sales charts (Huguenin, 2009).

In spite of the shifted price point, there are a number of benefits to software developers and distributors on the emerging mobile platform. First, changes in screen size allow for a very different experience, particularly with the development of tablets like the iPad (Hartley, 2010). Second, old content can be redeployed for additional profit. A number of developers are re-releasing games initially developed for PCs as mobile games (Economist, 2009). Third, development is cheaper because the game story and gameplay only need to be re-versioned for the new platform. This raises the potential for profit. Sega recently achieved sales of more than US $3 million for its game *Super Monkey Ball* (2001), which cost considerably less to develop than a Triple-A console game. Estimates suggest that iPhone apps can be developed for as little as US $5,000, though high-quality ones might cost more than US $150,000 (Rusak, 2009). Finally, the rise of app stores as an alternative distribution system means they can sidestep both traditional retail and the dominant software publishing sector.

The benefit to software developers is a significant one. Unlike traditional software publishing deals, which yield developers revenues of approximately 20 per cent of the software price tag, developing game apps allows higher cuts of the profits (Economist, 2009). For example, Apple leaves roughly 70 per cent of the sale of an app to the developer, an amount made more significant because they also avoid packaging and distribution costs (National Post, 2009). The Standard Developer Kit needed to create an app for Apple costs approximately $100, while development kits for consoles and for Nintendo or Sony handhelds come directly from the publisher and may cost more than US $2,000 (Rusak, 2009; Snow, 2009). Apple's estimates suggest that as many as 60 per cent of all people creating apps have never developed for Apple or other platforms previously (Beaumont, 2009). On the one hand, facilitating the process has allowed emerging programmers to make high profits. One programmer, Ethan Nicholas, who created the game *iShoot* (2008), is believed to have earned more than US $30,000 in one day, and more than US $1 million in a month of app sales (Rusak, 2009). On the other hand, it has resulted in a market flooded with apps, many of low quality. This has made it harder to make a profit on an app: unless a developer makes many apps, it is extremely rare and difficult to sell enough of a single app to make a truly sizeable gain (National Post, 2009; Rusak, 2009).

In spite of this, apps are drawing increasing attention from game makers and owners of media content. The development of apps for children is becoming particularly popular. Media groups including Nickelodeon, PBS Kids Sprout and Disney have all begun to invest in app development and developers (Rusak, 2009). Moreover, industry makers of consoles, particularly handhelds, are beginning to look more seriously at the area. In particular, Sony, which has struggled in recent years both in the console and handheld markets, has begun to develop a line-up of handhelds rumoured to include a smart phone in hopes of re-establishing its dominance in the hardware sector (Wakabayashi and Kane, 2010).

CASE STUDY: SONY

Japanese electronics giant Sony entered the video game industry as part of its push to expand into media content. Long before the company was known for video games, it was well regarded as an innovator in the consumer electronics business for its Walkman mobile personal audiocassette player, as well as its role in the production of computers, semiconductors and media recording. Starting with its purchase of CBS Records in 1987 and Columbia Pictures in 1989, the company began to seek ways to couple content with its extensive array of devices (Citron and Helm, 1991; Farhi, 1994). These purchases made Sony one of the largest media conglomerates, dominating the global media landscape (Wakabayashi, 2009). While the company had made video games for the company's existing consoles, its entrance into video game hardware was seen as a way to leverage its growing libraries of content with its well-regarded hardware (Harmon, 1993). Table 3.3. details Sony's major software and hardware franchises, as well as other key hardware developments.

Sony's entrance into video games began in 1994 with its first hardware console, the PlayStation. At that time the console market was dominated by Sega and Nintendo, but within three years, sales of the PlayStation in the US were outstripping rival platforms and were so good in the UK that Sony had to pull shipments from Europe and Japan to meet demand (Pescovitz, 1997; Turner, 1997). Much of this success owed to its ability to leverage a variety of technologies. With the 2000 release of the PlayStation 2 (PS2), Sony brought together these technologies, so that consoles could be used for video games but also to play CDs, DVDs and to access the Internet. In addition, the PS2 was backwards compatible, so that games from the previous system could be played on the new console (Grover *et al.*, 2005; Guth, 2002).

With the success of the PlayStation and the PlayStation 2, Sony was the dominant console manufacturer until the next generation of systems was released. It continued to leverage its wide technical capabilities into its consoles, allowing functions other than gameplay. In spite of its concentration, the hardware sector is volatile. As Figure 3.7 shows, the dominance Sony held in the console sector began to evaporate as new consoles were released. With the launch of seventh-generation consoles in 2006 – the Wii, the Xbox 360 and the PlayStation 3 (PS3) – Sony lost considerable market share. Though the PS3 console sold well, with more than 1 million sold within the first three weeks of release, its sales have lagged behind the competition (Leisure and Travel Week, 2009). This was partly due to the delayed release of the PS3, the last console of that generation. But the PS3 was also the most expensive to develop and the costliest to buy. Estimates suggest that the high-end model may have cost as much as US $800 to manufacture, and sold for US $499 (Graft, 2009). Despite

Year formed	**1933**
Headquarters	**Tokyo, Japan**
2012 Sales (US$m)	**78,912.0**
2012 Employees	**162,700**
Industry sector	**Hardware, software publisher**
Hardware and year introduced	
Consoles	PlayStation (PS) – 1994
	PlayStation 2 (PS2) – 2000
	PlayStation 3 (PS3) – 2006
Handhelds	PlayStation Portable (PSP) – 2004
	PlayStation Portable Go (PSP Go) – 2009
Non-game hardware	Trinitron monitors – 1968
	Betamax – 1975
	Walkman – 1979
	3.5" floppy disks/90 mm micro diskettes – 1983
	MSX (home computer system) – 1983
	CD (in collaboration with Phillips) – 1983
	Discman – 1984
	Handycam and Video8 format – 1984
	Minidisc – 1993
	Sony Dynamic Digital Sound (SDDS) – 1993
	Blu-ray DVD – 2003
Key software franchises	*Final Fantasy*
	Crash Bandicoot
	Gran Turismo
	God of War
	Spyro
	SOCOM
	Ratchet and Clank
	Twisted Metal

Table 3.3 Corporate Profile of Sony Corporation, 2012
Sources: Hoover's (2012); Mergent (2008).

its slow start, by 2009 the PS3 had begun to close the gap on its competitors. The cost of console production had dropped as had purchase price, pushing the console to 134 per cent sales growth (Media, 2009). By March 2010, Sony had sold 36 million PS3s worldwide, compared to 40 million Xbox 360s and 71 million Wiis (Kane and Wingfield, 2010).

Sony's success with the PS2 also allowed it to enter the handheld sector, taking on industry leader, Nintendo. In 2004, Sony released its PlayStation Portable (PSP) as a challenge to Nintendo's Gameboy series. Prior to this, Nintendo dominated the handheld sector, but Sony was able to break into the market.

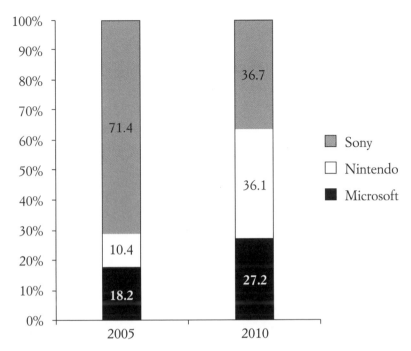

Figure 3.7 Percentage of Worldwide Console Market Share, 2005 and 2010
Source: VGChartz (2011a).

Figure 3.8 compares market share in handhelds by company from 2005 to 2010, indicating some gains by Sony. It is also worth noting that, while the proportions have changed, the number of handhelds manufactured by the various companies has gone down (VGChartz, 2011a). From the time of its launch, the PSP sold more than 55 million units, but this was considerably fewer than Sony had hoped and, as with the PS3, Sony suffered delays with the PSP, forcing the company to slash the number of units it shipped. (Wakabayashi and Kane, 2010). Released in 2009, its latest handheld, the PlayStation Portable Go (PSP GO), has suffered similar problems. With a price tag of around US $250 and requiring players to download costly games, sales have been slow (Media, 2009; Wakabayashi and Kane, 2010).

Perhaps the greatest competition to handhelds comes from the emerging smart phone market. Games for the PSP and PSP Go can cost as much as US $40, while most games for download on smart phones cost no more than US $5 (Media, 2009). While often more expensive than the PSP, many smart phones are multifunctional, with the ability to play movies, music and games (Gwinn, 2009). More problematically, many games for the PSP can also be purchased for the iPhone, prompting Sony to release simplified versions of many of its popular games, titled 'PlayStation Minis', and priced to compete with apps from the

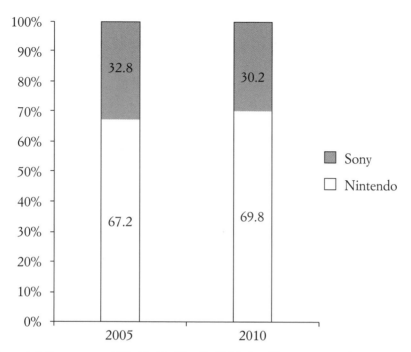

Figure 3.8 Percentage of Worldwide Handheld Market Share, 2005 and 2010
Source: VGChartz (2011a).

iTunes Store (Huguenin, 2009). While some analysts see the iPhone, iPad and iPod Touch as direct competition, chipping away at the established handheld sector, Sony has taken the view that these devices serve as a point of entry into more involved gaming (Kim, 2010; Rusak, 2009). Sony has attempted to enter into the production of smart phones, partnering with Ericsson in 2001. Despite prompt release of smart phones, the partnership struggled. Global shipments declined by more than 40 per cent in 2009, though the company released a smart phone of its own in April 2010 in hopes of addressing this (Wakabayashi and Kane, 2010).

Similarly, digital distribution poses an increasing challenge because it has the potential to disrupt relationships between developers and publishers within the industry. The success of outside companies has outpaced that of industry attempts. Like Microsoft and Nintendo, Sony has its own system for down-loading content but the PlayStation network has lagged behind the others, with 50 million subscribers by early 2010 (PR Newswire, 2010b). The threat of digital distribution changes the nature of the industry. As David Cole, an analyst at DFC Intelligence, sees it, 'video games are becoming more of a service [requiring the industry] to monetize users who will pay the most [while] getting more of those users who are willing to pay just a little ...' (Snider, 2010). Emerging technologies like digital distribution could rewrite industry timetables because distribution will be less tied to the major holidays for marketing, changing the game for developers and hardware manufacturers alike (Kane and Wingfield, 2010).

DIGITAL DISTRIBUTION

The ability to distribute content digitally is one of the key characteristics of the most recent hardware epoch, as discussed in Chapter 1. Its development may herald some challenges for the next epoch. The shift towards digital distribution owes much to the development of the casual game market described in Chapter 2. In 2001, revenues for Internet-related games in the US were only US $160 million, a fraction of the overall market for video games (Waldman, 2006). As casual games and games for mobile phones began to develop around 2004 and 2005, Internet companies like Yahoo! and AOL experimented with online game portals that worked across personal computer platforms and required only subscription fees and no software purchases (AP, 2005d). By 2005, North American Internet-related game revenues earned almost US $1 billion, with more than approximately a quarter of revenues coming from games distributed entirely digitally (Waldman, 2006). While these portals experienced some limited success, it wasn't until the seventh generation of consoles emerged, together with the advent of smart phones, that digital distribution began to have a real impact. From 2009 to 2010, digital distribution from game portals declined significantly, while consoles and smart phone distribution increased (Entertainment and Business Newsweekly, 2010).

The high cost of some of the consoles in the seventh generation, combined with high software costs for new game releases (typically between US $40–60) sparked resistance from consumers. Charging high prices for software allowed developers and distributors to minimise the impact of the high cost of producing Triple-A games (Economist, 2009). In contrast, digital downloads require minimal overhead, with no packaging, shipping and storage costs, easing the burden on consumers and improving sales margins for developers (Waldman, 2006). The impact of digitally distributed games is more significant in light of the small percentage of consumers who make use of them. Studies have shown that 80 per cent of gamers don't purchase games digitally and that the remaining 20 per cent who do so still obtain a majority of their games through retail stores (Entertainment and Business Newsweekly, 2010). Most of these purchases represent games in the casual market, challenging the industry to determine how to make use of digital distribution for other types of games (Bruno, 2009).

One of the first examples of a Triple-A game attempting to use digital distribution was *Grand Theft Auto IV* (2008), developed by Rockstar Games and distributed by Take-Two Interactive. This title allowed gamers to identify songs in its soundtrack that could then be purchased through Amazon.com (Bruno, 2009). Estimates suggest almost 700,000 users took advantage of the feature, though exact figures on the number of songs bought aren't available. Similarly, the *Rock Band* (2007) and *Guitar Hero* (2005) series of games have taken advantage of digital downloads to sell more than 50 million tracks for use within them. The *Rock Band* series, developed by MTV Games and distributed by Electronic Arts, maintains more than 500 songs sold digitally across hardware platforms, and it is estimated that an average of four songs are sold per user (Bruno, 2009).

According to Mike McGuire, a vice president at Gartner Research, digital distribution allows '[the major companies to extend] the value of the hardware platform' (Stelter, 2010). The major hardware distributors have taken notice, each developing its own online distribution system tied to its respective console platforms. Of the three hardware manufacturers, Microsoft's Xbox Live system has been the most successful despite having the highest subscription price (Shaw, 2010). In 2008, Sony's PlayStation network had more than 7 million users, while Microsoft's Xbox Live network had more than 10 million (Kane, 2008). Nintendo's Wii network, launched in 2008, has lagged behind Sony and Microsoft in terms of the number of original games available (Hillis, 2008).

Xbox Live was launched in 2002 and in four years had more than 1.5 billion hours of playtime (Waldman, 2006). Like the competing networks, Xbox Live allows users to access a range of media including games, film and television. Microsoft claimed the Xbox Live system regularly has more than 1 million users online at any time. In December 2009, the system registered 2.2 million simultaneous users (Stelter, 2010). Goldman Sachs has estimated that Microsoft earned more than €1.1 billion in sales in 2009, making Xbox Live a significant revenue source for the company, which at that time had only just begun to profit from console sales and was losing to Apple and Google in the mobile phone market (Irish Times, 2010). Xbox Live is an example of Microsoft's desire to transform its console into an entertainment hub, with executives arguing that the system is really a cable channel rather than a digital distribution platform (Stelter, 2010). In reality, however, it's actually a series of channels. Xbox Live Arcade, the portion focused on digitally distributed games, launched in 2004 and saw more than 6 million downloads in its first eighteen months of existence (Waldman, 2006). By 2009, Xbox Live had earned more than €1 billion sales, up from €800 million the previous year, with more than 25 million users paying an annual fee of US $5. In the first half of that fiscal year, which ended 30 June, subscriptions alone brought in US $600 million (Calgary Herald, 2010). More crucially, however, sales of movie and television programming beat subscription revenue for the first time (Irish Times, 2010).

Sony, Nintendo and Microsoft are working to update their networks in hopes of integrating new technologies and growing revenue. Sony aims to build its network into something similar to Apple's iTunes system, enabling any Sony device to download a range of content including video games (Wakabayashi, 2009). All three companies have begun to incorporate motion technology and are working towards the use of 3D gameplay and viewing in hopes of expanding how players use their hardware and networks (Kane and Wingfield, 2010). But digital distribution has the potential to shake up the existing structure, allowing major companies like Apple, not typically seen as a force in video games, to make their presence known, as well as allowing smaller players in the video game industry a chance to make their mark.

Companies like Valve, Mynga, Game Network, Gameloft and others are poised to take advantage of digital distribution and the casual game market (Kane and Wingfield, 2010). Software developer Valve, perhaps best known for its *Half-Life* (1998) series of games, was founded in 1996 (Valve, 2011). Table 3.4 details its major software franchises but information is limited because the company is privately owned. Valve has capitalised on a falling market for PC games. Its games are available in stores as well as via digital download. By 2010, downloadable PC games began to earn more than boxed games sold in stores, and estimates suggest that Valve controls between 50 to 70 per cent of the PC download market (Chiang, 2011). The company has been a champion of digital distribution since the launch of its online network, Steam, in 2007 (Sheffield, 2009). Steam has grown from 10 million subscribers

Year formed	1996
Headquarters	Bellevue, Washington, US
2012 Estimated sales (US$m)	8.9
2012 Estimated employees	50
Industry sector	Software developer, publisher
Digital distribution/publication platform	Steam
Key software franchises	*Counter Strike*
	Day of Defeat
	Half-Life
	Left for Dead
	Portal
	Team Fortress

Table 3.4 Corporate Profile of Valve Corporation, 2012

Sources: Business and Company Resource Center (2011b); Hoover's (2012); Valve (2011).

in 2008 to more than 30 million today, making it a serious rival to the networks run by Microsoft, Sony and Nintendo (Chiang, 2011; Sheffield, 2009; Valve, 2011). Valve's system allows users across PC and Mac platforms to purchase games, connect with each other and maintain the most up-to-date versions of the titles they've bought. The company has also begun to push into the living room, inking a deal with Sony to put a version of Steam on the PS3, as well as developing a 'big screen' version so the software can be played on televisions (Newman, 2011). Steam has become popular worldwide, translated into more than twenty languages and with content servers on six of the seven continents (Valve, 2011).

A second firm benefiting from digital distribution is Gameloft, founded in the UK in 1999 (PR Newswire, 2009). Gameloft has focused on making games for mobile devices and Facebook, often adapting existing games to work on these platforms. One of its first successes was *Gangstar: West Coast Hustle* (2009), a mobile application reminiscent of Rockstar's *Grand Theft Auto* series (Huguenin, 2009). As the smart phone market took off, Gameloft experienced tremendous profit growth. In the first six months of 2009, sales totalled €60.1 million, an increase of 20 per cent from the previous year. More than 95 per cent of those sales came from mobile gaming. Perhaps more importantly, Gameloft has drawn much of its revenue from Europe, with 38 per cent of its sales earned there (PR Newswire, 2009).

Should mobile games and digital distribution continue to proceed at the pace described, there is the potential for both the current industry structure to shift and for increased presence from developers outside the currently dominant regions. By allowing designers to develop games more cheaply and potentially to avoid the distribution networks established by the EA, Microsoft, Nintendo, Sony and the other major publishers, digital distribution could destabilise not just major players but an industry structure that has been stable for more than twenty years.

Digital distribution combined with mobile platforms has also opened up new markets and new potential for development, particularly in countries and regions mainstream games have not reached. One such example is Kenya, a country lacking the broadband infrastructure needed for the sixth- and seventh-generation consoles but which has significant mobile penetration (Moss, 2013). Similarly, game development in the Middle East has begun to blossom, relying on alternative payment methods made possible by mobile distribution (Lien, 2013). By circumventing broadband requirements, it is possible that an alternative to the mainstream industry may spring up in regions long ignored.

ARCADES

As video games have increasingly moved into the home, the public arcade has gone into decline in many countries (Vogel, 2004). In the US, there is little cross-ownership between the major hardware or software developers and the arcade sector. While figures for the sector are hard to come by because many arcades are privately owned, one example is particularly telling. Arcade profits at college campuses, where arcades were once hotspots, have fallen steadily since 2000. At the University of California's Berkeley campus, profits dropped from $400,000 a year to just over $50,000 while UCLA's arcade profits fell from roughly $700,000 to just under $250,000 per year by 2003 (Rooney, 2003). Now it is not uncommon for arcades to serve as test markets for popular games (McCallister, 2005). This is particularly true in countries outside the US where arcade games have maintained their popularity over the past decade. It is also worth nothing that, though arcade games make up only a small part of the current market, they do still bring in profits and have introduced a number of important franchises that have crossed over to the currently dominant console and PC platforms. Perhaps the best example can be seen in the *Mortal Kombat* series, introduced by Midway in 1992 as an arcade game, which now continues on various consoles.

In the US, however, arcades have fallen on rough times, caused in part by a negative reputation about their effects on players. Nolan Bushnell, who founded one of the first video game companies, Atari, and was one of the initial investors in Chuck E. Cheese Pizza Time Theaters, claims that video games have become about 'social isolation' (Richtel, 2005a). This has led Bushnell to open a series of restaurants called uWink with games available to play at the tables.

Bushnell's attempts to create a new family-friendly arcade experience are not the only vision for the future. Increasingly, attempts have been made to produce social spaces for older gamers. One example is Playdium.net, which opened in College Park, Maryland in early 2005. Here a local area network (LAN) was combined with first-person shooting games to connect players (Bowen, 2005). Playdium.net relied on tournament gaming for its profits, but players could also purchase game time in half-hour increments. This is similar to Chicago's Battletech Center, founded in 1990, which attracts players with its high-end video games, costing as much as $16 per hour for play (Nelson, 1990b). Playdium.net's founder hopes that, beyond the current business model, recruitment firms and government agencies may begin to use games as a way to screen applicants, a market Playdium could capitalise on (Bowen, 2005). Such schemes have been more successful in larger urban areas. Another business that draws on both Playdium's networked gaming model and on the high-end games featured at Battletech Center, is Dave and Buster's, an upscale dining and gaming establishment targeted at adults. With more than forty-four locations in the US and Mexico, Dave and Buster's adopted the franchise model of restaurants to create an alternative to the single-owned arcade (Dave and Buster's, 2005). An interesting spin on these business models can be found at Big City Gaming in Eugene, Oregon, where game rentals and sales were integrated with tournaments and in-store play (Nichols, 2005). This is a model that has become increasingly common among independent retailers, which often lose out to larger chain stores.

RETAIL

Though 2010 studies indicated that retailers have experienced only a small loss of sales due to digital distribution, the sector is watching how new distribution technologies develop. While studies suggest that as many as 80 per cent of game players don't purchase digitally, the rapid development of new platforms seems to show that a shift is coming (Entertainment and Business Newsweekly, 2010). Like the arcade sector, retail has remained largely unaffiliated with the major producers, with the exception of Nintendo, which opened a few retail stores whose success remains to be seen (Rosenbloom, 2005). As game rentals take off, it seems likely that the retail sector may receive more attention from hardware and software publishers in the near future.

Since 2005, the largest video game retailer in the US is GameStop, with more than 3,800 stores nationwide (New York Times, 2005a). Table 3.5 provides a corporate profile of GameStop. In 2005, GameStop alone accounted for $3.8 billion of sales in the US (Gamasutra, 2005c). This fuelled the company's expansion and in 2008 GameStop purchased French retailer Micromania, boosting its presence in the growing European market (Martin, 2008). The move into the global market changed GameStop's revenue profile. By 2009, the company was drawing increasing amounts of its revenue from outside the US: 73 per cent of revenues were from the US, while Europe accounted for almost 15 per cent and Australia for nearly 6 per cent. (Matthews, 2009b). Earlier, the acquisition of rival Electronics Boutique had made GameStop the dominant chain in the US with roughly a quarter of US sales, its closest competitor Wal-Mart accounting for another 25 per cent (New York Times, 2005a).

This heavy concentration means that the large number of small independent retailers have a hard time affecting the market in any substantial fashion. Figure 3.9 breaks down the market share by retailer of game sales in the US for the year 2008. GameStop has a significant market share but is also the only retail outlet focused solely on video games. For most of the merchants listed, games only represent a part of their business, so that they can be treated as loss-leaders for other products if so desired. This level of concentration has allowed

Year formed	1996
Headquarters	Grapevine, Texas, US
2012 Sales (US$m)	9,550.5
2012 Employees	71,000
Industry sector	**Retailer**
Key stores	GameStop
	Micromania
	EB Games
	Electronics Boutique
Other interests	*Game Informer* (magazine)

Table 3.5 Corporate Profile of GameStop, Corp., 2012
Sources: Business and Company Resource Center (2011a); Matthews (2009a, 2009b).

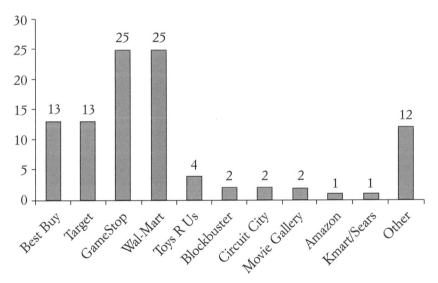

Figure 3.9 Percentage of US Video Game Retail Market Share, 2008
Source: Market Share Reporter (2009c).

GameStop to corner the market for used game sales in the US, with one estimate for 2007 giving the company control of 80 per cent of used sales (Market Share Reporter, 2009d). One of the contradictions in the market is that used games have been blamed for an 8 per cent decline in sales of new titles, prompting the industry to devise ways to curb reselling, for instance, by including placing protection on the software to limit playability after initial use and charging additional fees (Snider, 2010).

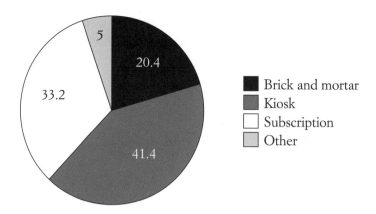

Figure 3.10 International Video Game, DVD and Video Rental Markets Share, 2011
Source: IBIS World (2012).

Video game rentals are also heavily concentrated. It is not uncommon for stores that rent films to also rent games and, as digital distribution becomes more of a reality, both film and games are seen as key components in its expansion. Figure 3.10 breaks down the video game, DVD and video rental market in the US during 2011. While the production of games is considerably different, the distribution arm of the business, including rentals, functions similarly to that of the Hollywood film industry. Thus, it is important to note that these companies are the same ones that dominate video and DVD rentals in the US (Wasko, 2003). As the chart also indicates, brick-and-mortar stores face tremendous challenges from alternative rental options, including mail subscription services like GameFly and Netflix as well as from kiosk rental services like RedBox. A more detailed analysis of the ties between the video game and film industry can be found in Chapter 4. It should also be noted that game rentals happen exclusively on console and handheld platforms. Not surprisingly, the consoles that dominate the market tend to dominate rentals. This has helped Sony take advantage of rentals to a much greater degree than the company's share of the most recent platforms globally, suggesting that there is still a significant use and market for products developed for earlier platforms, particularly the PlayStation 2, which continues to sell well globally. Such entrenched usage indicates a more significant penetration than the industry has seen in previous decades. Typically, studies focus on market penetration of the most current generation, but if, as is the case with the PS2, they continue to be popular after newer platforms have been released, this additional market must be factored in. The retail sector, however, remains the public's primary view of how the video game industry works, drawing consistent scrutiny from parent groups for selling practices seen as putting questionable content into the hands of children. This has led to retailers enforcing strict controls on their own employees, such as BestBuy's 2005 policy promising harsh sanctions for employees who fail to check the ages of shoppers (Ivry, 2005).

CONCLUSION

The modern era of video game production has resulted in distinct developments in hardware. As consoles have become more popular, they have begun to influence computing in other ways, and have been adapted for a variety of purposes. As this has happened, consoles have become a driving force in computer manufacture and as platforms for different media experiences. For the industry, hardware manufacture has been the costliest element, and sales of software have been used to defray these costs. Traditional wisdom is that the hardware sector can only support three major consoles at a time. But the reality is much more complicated, with a variety of consoles, handhelds and personal devices supporting gameplay. Moreover, as the rise of the Zeebo console suggests, part of the reason for this limit is that the industry has restricted itself by concentrating on the North American, European and Japanese markets. It remains to be seen how the Zeebo will fare against consoles from companies that have the significant advantage of also owning three of the main software distributors.

In this context, the hardware market is heavily concentrated, though volatile, with the major players jockeying for the lead from one console generation to the next. Success relies partly on continued innovation but, as console prices rise, the market seems to be reaching the limits of what consumers are willing to pay. This may account for some of the continued loyalty to the PS2. Indeed, new technologies – particularly mobile phones and digital distribution – are poised to erode the dominance of the three major consoles in no small part because they are characterised by cheaper software, more casual play and easier use than many of the existing consoles. The rise of cell phones and tablets as game platforms highlights two major challenges for the industry. First, they are forms of hardware that the major console players have little control over. Second, they represent a shift in terms of distribution from the Convergence Epoch into the Networked Epoch, allowing consumers to download directly rather than purchase video games. Digital distribution poses a challenge for retailers as well. The retail sector is already concentrated, with certain chains controlling sales of both new and used software sales. Increasing sales of used games have become something of a problem for the industry, with consumers reacting negatively to rising prices for software. Digital distribution platforms like Valve's Steam are seen as one way to cut into this concentrated market, and in response the major console manufacturers have all begun to expand their own networks to try to take advantage of this trend.

As will be seen in Chapter 4, new platforms like cell phones also pose a threat as they compete for key components from one of the video game industry's key allies: microchip manufacturers. As game products become more popular globally, the industry has sought to formalise ties with other major cultural industries, particularly the film, recorded music and sports entertainment industries. These relationships, as well as ties to microchip and computer manufacturers, will be explored in more detail in the next chapter.

4

Video Games and Other Related Industries

As discussed previously, the relationship between video games and other communication industries has a long history that is often neglected. Chapters 2 and 3 provided snapshots of the structure of the modern industry, which is increasingly engaged in more than just the creation of video game products. Because most revenues come from the sale of software – commodities that only sometimes result in recurring purchases – the revenue has had to be maximised in other ways. By fostering ties with other communication companies, the industry hopes to capitalise on both their established brands and those that have thrived elsewhere in the cultural marketplace.

Video game audiences have grown, consistently drawing consumers from other media. One study suggested that 52 per cent of gamers increased the time they spent playing by watching less television (Donaton, 2003). Younger audiences, in particular, are spending more time on video games. One study found that children's media access had grown considerably, with the average child having access to VCRs, DVDs, video games and computers in their bedroom (Gutnick *et al.*, 2011). The increasing ubiquity of equipment suitable to play video games on, combined with the growing audiences discussed earlier, has resulted in some experts predicting that they will start to be used in political campaigns (Foster, 2004b). Indeed, at least one campaign – Howard Dean's bid for the US presidency – has already taken this step (Erard, 2004).

The ability of video games to reach diverse audiences in diverse ways has meant that cross-industry promotion is not one-directional; video games are increasingly a viable platform for maximising revenue from a variety of media streams. Though historically, other cultural industries have had problems incorporating games into their strategies, a number of ways have been found around these problems, instituting video games as a vital part of the modern media landscape. They are also attracting more audience time, making them a competitor for audiences and advertising revenue. Increasingly, this has meant that video games must be incorporated into the synergistic plans of a wide range of media products and brands, including film, television and recorded music. Not surprisingly, games have become a driver of high-tech innovation as well.

This chapter examines the ties between video games and other media industries. It begins by discussing the importance of contextualising the hardware industry with regard to microchip processing, then debates the importance of licensing and franchises on the software side. It then analyses the interplay between the video game industry and film, television, recorded music and publishing industries. In addition, it examines the connection between video games and sports, one of the most profitable relationships the industry has developed. It concludes by looking at the growing importance of ties with advertising, including the development of in-game advertisements and of custom games.

VIDEO GAMES AND THE MICROCHIP INDUSTRY

As seen in Chapter 1, the modern video game industry owes much of its development to the computer industry. While frequently seen as a subset of the computer and microchip business in the 1970s, today video games have become a separate industry and an important driver for the global production of computers and, in particular, microchips. The chips in game consoles are typically more expensive than those in most PCs, in part because they need to offer better graphics and processing speed (FinancialWire, 2007). The demand for computer memory chips grew so rapidly, in part due to the surge in video game popularity, that global prices for chips surged 38 per cent in the year 2006 and have continued to rise; the demand for chips for consoles and smart phones is so intense that major PC manufacturers have begun to experience shortages (Cho, 2006). In fact, game consoles have become competitors in the global marketplace, owing to their advanced graphical and processing capabilities. Estimates suggest that by 2007, video games accounted for as much as one-quarter of the world's general use of computer storage (Hilbert and López, 2011). Proof of this can be seen in examples such as the US Air Force's purchase of more than 2,000 Sony PlayStation 3 consoles (Betts, 2009). Chip manufacturer nVidia has sold its graphics processors to workers in oil and natural gas, cancer researchers and designers (Schmerken, 2008). Similarly, the graphics cards typically associated with games consoles are being used in financial trading as well, where their advanced processing power has helped to calculate the complex analytics needed for options and derivatives (Schmerken, 2008).

The centrality of graphical capability to video games makes them a logical first place to start in understanding the relationship between the industry and chip manufacturers. As Figure 4.1 shows, this sector of chip manufacture has maintained heavy concentration for several years, with three manufacturers dominating the market: ATI, nVidia and Intel. This concentration only increases

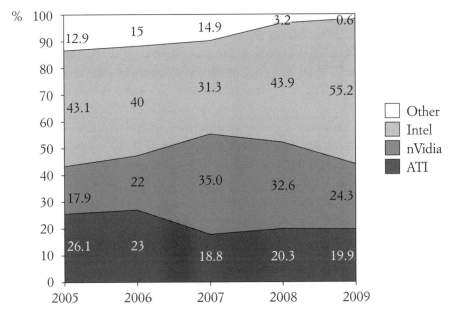

Figure 4.1 Global Market Share of Graphics Chips Manufacturers, 2005–9
Sources: Market Share Reporter (2005a, 2007, 2008b, 2009a, 2010a).

when applied to game consoles. For the seventh-generation consoles, only two of those companies manufacture the graphics chips. ATI manufactures the chips for the Xbox 360 and the Nintendo Wii, while nVidia is the manufacturer for Sony PlayStation3 (Alpert, 2003). Intel has retained its lead in the sector because it's employed in so many personal computers, though it is less popular for gaming.

There is, of course, another type of chip in both personal computers and video games. These chips deal with processing and memory, and they have come to constitute a crucial battle in the manufacturing industry as well as a key supply point issue in hardware manufacture. With the seventh generation of consoles, one company has come to dominate this sector: IBM, which established a beach-head with console manufacturers by helping Sony make its Cell processor. The Cell processor was a risky development, shared between IBM, Sony and Toshiba (Takahashi, 2004). Estimates indicate that Sony alone invested more than US $1.7 billion in development for the PlayStation 3 (International Herald Tribune, 2007). The chip was designed to process information in a different and faster way than existing chips. Cell chips are said to have approximately 250 million transistors, roughly double the number of the Intel's Pentium 4 chips (Turley, 2006). The cost is assumed to be worth it, as the chips can be used for a range of products beyond games, generating interest from the defence, medical and entertainment sectors in particular (Bergstein and AP, 2006b). The costs would also eventually be mitigated as the companies reached economies of scale in manufacturing, which has helped to bring the high cost of the PlayStation 3 down (Guth, 2005). By manufacturing its own chip, Sony also insulated itself from turns in the market, allowing it to minimise losses and potential workflow stoppages.

After working with Sony for more than two years, IBM won the contract for Microsoft's consoles away from Intel, a major coup in the sector (Bergstein and AP, 2006b; Sloan, 2009). The ability to control costs and production led both Microsoft and Nintendo to also seek a manufacturer for processing chips for their consoles. For Microsoft, supply chain problems and unpredictable chip costs during the manufacture of its first console, the Xbox, made it determined to control manufacture of its processing chips. The original Xbox relied on Intel's Pentium 3 chips, but troubles with supply were a large factor in the nearly US $4 billion in losses from 2001 to 2005 (Guth, 2005). In fact, the loss is estimated to have been as much as US $125 on each Xbox because of high component costs. Microsoft talked with Intel and nVidia as well, but ultimately chose IBM in order to maintain ownership of the chip design (Takahashi, 2005). Nintendo's manufacturer landed with IBM as well, giving the corporation a monopoly in the console sector. This partnership brought the chip manufacturing segment back from severe losses. In 2003, it is estimated that IBM's microelectronics division lost more than US $250 million. In contrast, following the launch of the seventh-generation consoles in 2006, its estimated earnings from chip sales were more than $3.7 billion (Bergstein and AP, 2006b).

The chips IBM designed for each console have individualised capabilities, though each shares a common multiple-core processor, facilitating faster processing speeds (Bergstein and AP, 2006b). For example, the chip for the Wii not only vouchsafed faster processing but also a reduction in energy consumption, a feature becoming more important as console energy use has come under fire (Knight-Ridder, 2006; Rahim, 2010). Based on the successes with consoles, IBM invested heavily in chip manufacture, sinking almost US $3 billion into its facilities alone. In addition, it formed a new 'technology collaboration solutions' unit, which is expected to become one of the company's key profit makers (Bergstein and AP, 2006b). Future chips are expected to be even more complex than the Cell and might contain more than 1,000 elements. IBM's ability and focus on more efficient power consumption and flexibility in its chips have been crucial to extending its manufacturing alliances (Markoff, 2006).

LICENSING AND FRANCHISES

As video game audiences and revenues have grown, one strategy that has reaped rewards is the use of licensed products. Licensed products tend to command big audiences, and the industry has found that drawing on concepts with a proven track record in other media increases the likelihood of profitability (Levine, 2005c). The shining example of this is the *Super Mario* franchise. By 2002, the *Mario* franchise had grossed more than US $7 billion globally (Guth and Tran, 2002). With the release of the seventh generation of consoles starting in 2006, more *Mario*-themed games added to this total. Estimates suggest that the franchise has earned US $2.3 billion in the US alone (Vargas, 2006). When the first *Mario* games were released in the early 1980s, video game franchises were rare, forcing developers to look outside the industry for licensed content. However, in the modern industry, successful franchises are becoming more common, serving as key entry points to consoles. Indeed, the most popular games become franchises of their own, extending their reach into other media, particularly into film and publishing. Table 4.1 details the bestselling franchises in 2007. It is worth noting that, while a number of franchises occurred early on, most develop following the re-emergence of video games around 1990 discussed in Chapter 1. Finally, it's important to note that, of all those franchises, only two hail from ties external to the industry – *Madden NFL* (1989) and *James Bond* – suggesting that there may be a limit to the viability of licensed content from other media.

It should not be surprising that most of these game franchises have found ways to move into other media, but that they have remained under the control of the big players in video games. Nintendo, Electronic Arts and Sony account for almost half of the franchises on the list. The release of updated numbers would almost certainly see Microsoft take up positions on the list as well. The longevity of these franchises provides additional evidence for anyone needing proof of the longevity of game audiences or the cultural relevance of content. The *James Bond* franchise also offers an interesting example of game development. The first Bond-themed game was released in 1983, but since then, the Bond property has been licensed to a number of developers and across platforms. Such an approach is common with properties adapted from outside the industry, though it is not always used. Some companies – most notably LucasArts – have experimented with licensing their properties to a single platform (LucasArts, 2002). Such console exclusivity is rare for outside content; however, franchises which have emerged within the industry are much more likely to follow this model. Industry concentration reinforces this tendency and, because the library of video game content is still relatively limited in comparison to film, television and recorded music, the industry often seeks out proven franchises and ways of adapting that content. Film has proven to be the most notable example of this (Holson, 2005a).

Franchise	Publisher with rights, 2008	Millions of units sold, 2007	Year of game's first appearance
Mario	Nintendo	193	1981 (as Jumpman in *Donkey Kong*) 1983 (*Mario Bros.*)
Pokémon	Nintendo	155	1996
Final Fantasy	Square Enix	68	1987
Madden NFL	Electronic Arts	56	1989
The Sims	Electronic Arts	54	1989 (*SimCity*) 2000 (*The Sims*)
Grand Theft Auto	Rockstar	50	1997
Donkey Kong	Nintendo	48	1981
The Legend of Zelda	Nintendo	47	1986
Sonic the Hedgehog	Sega	44	1991
Gran Turismo	Sony	44	1998
Lineage	NCSoft	43	1998
Dragon Quest	Square Enix	41	1986
Crash Bandicoot	Sony/Vivendi	34	1996
James Bond	Various	30	1983
Tomb Raider	Eidos Interactive	30	1996
Mega Man	Capcom	26	1987
Command and Conquer	Electronic Arts	25	1995
Street Fighter	Capcom	25	1987
Mortal Kombat	Midway	20	1992

Table 4.1 Bestselling Game Franchises, 2007
Source: Gamasutra (2007).

Other media industries are partly attracted to video games because of the high profit margins they can yield. The typical profit margin for a Hollywood studio production is around 10 per cent, while game makers can expect an average profit margin of 15 per cent on most games and, if they do particularly well, the bestselling games may make as much as a 25 per cent profit margin (Grover *et al.*, 2005). This suggests that film makers can make significantly more money on games than films, and the incorporation of games into a broader strategy of synergy might bring even more impressive profit margins (Fritz, 2004). For example, the first movies in the *Harry Potter* franchise delivered more than US $100 million in licensed product sales (Elkin, 2003). In some cases, the licensing deals might happen with little input from the video game industry. One example of this is EA's use of the Aston Martin in its James Bond game, part of a previous licensing deal between MGM Studios and the auto manufacturer, which suggests that probably neither publishers of any of the games nor developers saw any direct profit from the deal (Elkin, 2002c). Increasing licensing has been beneficial for Hollywood, in particular. In 2002, Disney reported more than US $13 billion in licensing while Time Warner drew almost US $6 billion (Elkin, 2003). Reliance on licensing is not without its perils, as players in particular have proven wary. According to a study by the ESA, more than one-third of all gamers would prefer to see fewer licensed games (Roch, 2004).

GAMES AND FILM

In spite of the audience's stated ambivalence to video games, other media industries have realised their considerable potential; none more so than the film industry. Indeed, both game and film companies have begun to recognise the possibilities of cross-industry licensing. Cross-industry licensing deals have been seen as an important way of growing audiences and deepening relations with established ones (Pereira, 2002a). In particular, the growth of the female audience has been a focal point for these deals (O'Connor, 2003). This can also be seen in which video games have succeeded as franchises and bestsellers. Table 4.2 compares the bestselling franchises in the video game industry from 2005 to 2011. As it shows, the most successful have traditionally been targeted towards younger audiences. Most of the games represented in the chart are targeted towards either family play or children specifically. It's also worth noting that, while Nintendo has come to dominate the bestselling franchises, it's done so across a variety of platforms.

Based on worldwide sales					
2005			2011		
Game title	Units sold	Publisher	Game title	Units sold	Publisher
Super Mario Bros.	40	Nintendo	Wii Play	78.68	Nintendo
Tetris	33	Nintendo	Super MarioBros.	40.24	Nintendo
Super Mario Bros. 3	18	Nintendo	Pokémon Blue Red/Green/ Version	31.37	Nintendo
Super Mario World	17	Nintendo	Tetris	30.26	Nintendo
Super Mario Land	14	Nintendo	Mario Kart Wii	29.76	Nintendo
Super Mario 64	11	Nintendo	Wii Sports Resort	29.41	Nintendo
Super Mario Bros. 2	10	Nintendo	Wii Play	28.43	Nintendo
The Sims	10	EA	Duck Hunt	28.31	Nintendo
Grand Theft Auto: Vice City	8	Rockstar	New Super Mario Bros.	26.41	Nintendo
Harry Potter and the Sorcerer's Stone	8	EA	Nintendogs	24.29	Nintendo

Table 4.2 Comparison of Bestselling Video Games, 2005 and 2011
Sources: Market Share Reporter (2005a); Nichols (2008); VGChartz (2011b).

While most of the 2011 series of games are for the Wii, *Super Mario Bros.*, *Duck Hunt* (1984), *Tetris* (1989) and the *Pokémon* (1996) games were all handheld. Nintendo has worked to maintain a family-friendly vibe for its platforms, so the fact that these games would centre around its platforms is less surprising. What may be surprising, particularly in light of the coverage games like *Halo 3* (2007) and the various *World of Warcraft* have received, is that by the end of 2011, those games had only sold 11.39 million and 15.75 million units, respectively (VGChartz, 2011b). In part, this would seem to underscore the growing importance of casual games, which share a similar sense of family appropriateness and ease of play with many of the bestselling games. However, only the *Mario* and *Pokémon* games have gone on to much licensing success.

For most of their history, the film industry has been the stronger partner in terms of licensing. This resulted in lopsided licensing deals. Typically, when film studios license rights to a property for video game development, they have expected from US $3 to 5 million up front, plus up to 9 per cent of the game's profits (Grover *et al.*, 2005). Because video games typically cost much more than even the most inexpensive theatre ticket, a game must be extremely popular in order to be as profitable as a Hollywood film (Fritz, 2004). The surge in game popularity, however, has begun to turn the tables. Both *Halo 3* and *Grand Theft Auto: San Andreas* debuted with stronger sales than major movies released the same week, signalling a sea-shift in the power dynamics between the two industries (Fritz, 2004). *Halo 3*, for example, generated more than US $125 million in its first weekend while hit move *The Incredibles* (2004) earned only US $70 million in the same weekend (Gentile, 2005b). An even bigger indicator, however, came when Microsoft decided to forgo working with Hollywood in producing a film based on its hit franchise *Halo* (Brodesser and Fritz, 2005).

Of course, the ties between the two industries have existed since the earliest days of video games. The first major game companies all viewed themselves as expanding on the movie business and on the possibilities of film itself. Moreover, Hollywood invested in video games and experimented with licensing movies to games in the mid-1970s. As noted previously, the failure of the video game based on the hit movie *E.T. The Extra Terrestrial* is often used to mark the first decline of video games in the early 1980s and was one of the reasons that Warner Communications, one of the first major entertainment companies to explore video games, pulled out of the business (Kent, 2001). The 1984 failure of the *E.T.* game resulted in 5 million cartridges being sent to a landfill, representing the biggest failure of the Hollywood/video game attempts at crossover (Grover *et al.*, 2005).

By the 1990s, Hollywood was again interested in games. Companies like Dreamworks SKG, Time Warner and Disney, as well as other media giants,

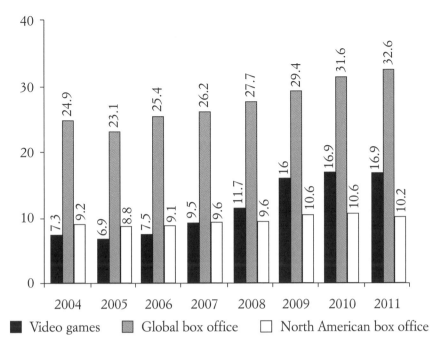

Figure 4.2 Comparison of US Video Game Software Sales with North American and
          Global Film Box-office Receipts in Millions of US Dollars, 2004–11
Sources: Box Office Mojo (2011b); ESA (2009, 2011, 2012); MPAA (2009, 2012).

began to build their own game development units. Ultimately, these attempts
faltered due to high development costs and Hollywood's limited vision of video
games' potential (Holson, 2005a). One comparison frequently made to justify
the importance of video games is that between games sales and Hollywood box
office. Figure 4.2 compares the revenues from software sales and Hollywood's
North American and global box office from 2004 to 2011. As it shows, even as
video games have become more popular, they have taken a long time to surpass
Hollywood box office. Moreover, while US sales have surpassed the North
American box office, they have not yet overtaken global box office sales. It's also
worth noting that this comparison ignores all of the other revenue streams
Hollywood film makers may be able to take advantage of, such as DVD sales,
which have become increasingly important to the industry's revenue streams. It
also doesn't look at the global picture. It does suggest, however, that video
games could indeed be a serious rival for entertainment dollars.

By the year 2000, however, many companies including Fox and DreamWorks
had all but abandoned their game units, preferring instead to license their
content and allow the video game industry to risk the high development costs
(Grover *et al.*, 2005). In part this owed to slower game sales growth, but also to

distinct differences in development culture between the two industries. As will be seen in the next chapter, labour in the video game industry owes more to the computing industry than to Hollywood. This period saw the first major in-roads for licensing games into films. Starting in 2003, a third foray into video games began, with Hollywood majors restructuring their franchise models to facilitate game development of various properties from their extensive libraries of content (Diamante, 2005). Rather than focusing on acquiring games companies, Hollywood began to enter more strict licensing arrangements and to develop units for particular types of film development, particularly, animations, which were seen as especially suitable properties. Table 4.3 details movies based on video games during this period. Perhaps surprisingly, only three of the bestselling game franchises of all time are represented there, though many franchises listed in a previous chart have also had movies based on them. Moreover, unlike the bestselling video games of all time, most of the films and the games they represent are less family friendly. In part, this may owe to higher ticket prices pushing more recent movies, which have been less family-oriented, but it may also suggest that different types of game content might translate more readily into profitable films.

In spite of the high profitability of video games, having even surpassed US box-office sales, Hollywood is still the dominant party (Diamante, 2005). Not only does the film industry have a more extensive catalogue, but it also has more avenues to ensure profit from a concept. Marketing the idea for a film or a video game through all film's ancillary markets, is a much greater guarantee of profitability than concentrating on a single commodity. In contrast, game revenues depend primarily on sales with a small amount coming from rentals. Moreover, games typically have a shelf life of roughly three months, unlike films that can be released and re-released over a longer period (Grover *et al.*, 2005). Increasingly, video games and film are becoming an integral part of both industries' synergy strategy in which successful pieces of intellectual property – characters, concepts, stories – are adapted for a variety of media forms. For example, in 2002 more than 100 games were based on movies and video releases (Tramain, 2002). In 2005, when Sony scheduled the launch of its PlayStation Portable platform, United International Pictures released a number of movies especially formatted for the device, including *8 Mile* (2002), *The Bourne Supremacy* (2004) and *The Mummy Returns* (2001) (Dawtrey, 2005). While typically video games have come out on or close to the release date of popular movies, in order to benefit from the film marketing, the launch is increasingly being pushed back to coincide with DVD releases and other film promotions or to extend the brand (Elkin, 2002c; 'Video Games Go to the Movies', 2003). Perhaps the best examples of this include games based on long-running, major franchises like the LucasArts'

Film title	Film studio	Adjusted box-office gross in millions of US dollars	Game publisher	Game release date (if franchise, first game's release used)
Lara Croft, Tomb Raider [West, S. 2001, US]	Paramount	131.2	Eidos	1996
Prince of Persia: The Sands of Time [Newell, M., 2010, US]	Buena Vista	90.6	Ubisoft	1989
Pokémon: The First Movie [Haigney, M. and Yuyama, K., 1998, Japan]	Warner Bros.	85.7	Nintendo	1998
Mortal Kombat [Anderson, P., 1995, US]	New Line	70.5	Midway	1992
Lara Croft, Tomb Raider: The Cradle of Life [de Bont, J., 2003, US]	Paramount	65.6	Eidos	1996
Resident Evil: Afterlife [Anderson, P., 2010, Germany]	Screen Gems	60.1	Capcom	1996
Resident Evil: Apocalypse [Witt, A., 2004, Germany]	Screen Gems	51.2	Capcom	1996
Resident Evil: Extinction [Mulcahy, R., 2007, France]	Screen Gems	50.6	Capcom	1996
Silent Hill [Gans, C., 2006, US]	Sony	47.0	Konami	1999
Pokémon: The Movie 2000 [Haigney, M. and Yuyama, K., 1999, Japan]	Warner Bros.	43.6	Nintendo	1998

Table 4.3 Top Ten Movies Based on Video Games by Gross, 2011
Source: Box Office Mojo (2011a).

franchise *Star Wars* (1977–), *The Lord of the Rings* (2001–3). Struggling franchises have also toyed with this while the film *Superman Returns* (2006) was a summer blockbuster, the game based on the film didn't come out until the Christmas buying season of the same year.

Perhaps because of a failure to understand how to integrate with video games, or because of their volatile early years, Hollywood has been cautious in its dealings with the gaming world even during its successful periods. Both Viacom, the parent company of Paramount Pictures, and Time Warner have become more active in exploring games. Viacom was noted as having interest in Midway games prior to the latter's 2009 bankruptcy, but Time Warner, which purchased its developer

Monolith in 2004, ultimately bought most of Midway's properties for $33 million (Thorsen, 2009; USA Today, 2004). Time Warner has since acquired a number of other studios, held under their Warner Bros. Interactive Entertainment division. These include Rocksteady Studios in England, NetherWorld Studios in Chicago, Snowbird Studios in Kirkland, Washington and Turbine Studios in Needham, Massachusetts (Time Warner, 2011). But it is the Walt Disney Corporation that has taken the most decisive steps. Table 4.4 provides an overview of Disney's interests in the video games industry as well as its other media holdings. In 2005,

Year formed	**1933**
Headquarters	**Burbank, California, US**
2011 Sales (US$m)	**4,807**
2011 Employees	**156,000**
Industry sector	**Software developer, publisher**
Key software and other franchises	*Mickey Mouse*
	Donald Duck
	Marvel Comics properties including The Avengers, Spider-Man, The X-Men, Iron Man, Thor and The Hulk
	Kim Possible
	Disney Princesses
	The Little Mermaid
	Pirates of the Caribbean
	Aladdin
	High School Musical
	Hannah Montana
	Hercules
	In the Night Kitchen
	Kingdom Hearts
	Spectrobe
	The Cheetah Girls
	The Lion King
	The Chronicles of Narnia
	Winnie the Pooh
	Tron
	Who Framed Roger Rabbit?
	Where the Wild Things Are
	Who Wants to Be a Millionaire?
	Turok
	Desperate Housewives
	Chicken Little
	Toontown
	The Muppets

Table 4.4 Corporate Profile of the Walt Disney Company, 2012

Table 4.4 *continued*

Media properties	ABC Television
	ESPN
	The Disney Channel
	Disney/Pixar Films
	Touchstone Pictures
	Hollywood Pictures
	Miramax Pictures
	Walt Disney Records
	Hollywood Records
	Lyric Records
	Radio Disney
	SOAPnet
Studios owned	Propaganda Games – Vancouver, British Columbia, Canada
	Avalanche Software – Salt Lake City, Utah, US
	Climax Racing – Venice, California, US
	Q Entertainment – Tokyo, Japan
	Junction Point Studios – Austin, Texas, US
	Gamestar – Shanghai, China
	Disney Interactive Studios Canada – Toronto, Ontario, Canada
	Disney Interactive Studios Latin America – Sao Paulo, Brazil
	Disney Interactive Studios UK – London, UK
	Disney Interactive Studios Japan – Tokyo, Japan
	Disney Interactive Studios Asia Pacific – Hong Kong

Profile

Disney, the world's second largest media company behind Time Warner, has once again entered the video game market, in spite of previous difficulties. The company renamed its studios in 2007. Disney Interactive Studios, formerly known as Buena Vista Games, has begun attempts to use the company's extensive catalogue of properties to corner the children's game market. Because the company's primary market is children, its products have been focused at platforms that are particularly child-friendly: personal computers and handhelds (Hoover's, 2008). The company continues, however, to license Disney intellectual property to other developers as well as working in conjunction with other developers and publishers. One notable example is *Kingdom Hearts*, produced with rival Square-Enix and with THQ for Disney/Pixar films (Disney, 2007c). This division operates from Glendale, California, but the company has operations in Brazil, Canada, the UK, Hong Kong and Japan as well (Disney, 2007c). Through the Disney board of directors the company has ties to Apple Computers, Proctor and Gamble, Edison International, Boeing, Cisco Systems, Estée Lauder Companies Inc., Sears, YUM! Brands, ImpreMedia, LLC, the Seagram Company, Ltd, and Starbucks, among other companies (Disney, 2007a, 2007b).

Sources: Hoover's (2012); Mergent (2008).

Disney purchased Vancouver studio Avalanche and scored a coup by luring an entire development team away from game giant Electronic Arts (Fritz, 2005c). Since that time and thanks to the rise of the casual games movement, Disney has increased its interests in the game industry with studios in Canada, Brazil, England, Japan and Hong Kong, among other locations (Disney, 2012). Following the success of its *Toontown* game which was launched online in 2001, Disney began to explore the online and multiplayer game space, developing a game based on its *Pirates of the Caribbean* franchise, which was launched in 2005 (Fritz, 2005d). Disney's 2004 acquisition of Jim Henson's Muppet properties, combined with its 2009 acquisition of Marvel Comics, has boosted the number of licensable properties it has available to explore and integrate into its games and other media properties (Davis, 2009; Grant and Orwall, 2004). Such acquisitions allow the company to reverse its traditional film-to-video game model, enabling games to drive content in other media (Fritz, 2005c). The addition of Marvel also addresses a key need. Through 2005, 90 per cent of the Disney's video game revenue was based on games sold to children, while the industry trend was moving in the opposite direction (Marr, 2005). One of the advantages of the Marvel properties – particularly the recent films – has been the much broader audience they have drawn (License!, 2005). Should Disney's games succeed in tapping into these broader audiences with games, they'll be a force to be reckoned with, potentially rivalling companies like Electronic Arts.

Perhaps due to the concentrated nature of the video game industry, other film companies have been slower to get involved, though some of them have explored the possible licensing opportunities. Prior to Midway's bankruptcy, Paramount licensed a number of its properties for development. In 2003, the trend came to the fore, with several films, including *Resident Evil* (2002), *House of the Dead* (2003) and *Lara Croft, Tomb Raider: The Cradle of Life* (2003) all being released (Herold, 2003a). Because of the popularity of video game properties, producer Uwe Boll was able to raise more than US $47 million to invest in films based on the *BloodRayne* (2002) and *Dungeon Siege* (2002) games (Harris, 2004). Though the films – *BloodRayne* (2005) and *In The Name of the King* (2006) – didn't earn back the investment at the box office, the trend for films based on games hasn't slowed. Part of the attraction would seem to be that these games fit the Hollywood action formula, which has been particularly successful internationally (Wasko, 2003). In contrast, it is surprising that the Hollywood companies perhaps most suited to working with video games – animation studios – have tended to sign outside licensing deals rather than create their own games based on their properties. Historically, animation studios like Pixar, DreamWorks Animation and Fox Animation Studios have all relied on outside game production (Fritz, 2005i).

Animation – particularly children's entertainment franchises – has become one of the prime development areas. After its success with the film *Finding Nemo* (2003) and the game based upon the film, Pixar signed one of the first deals between an animation studio and a game publisher, THQ (THQ, 2004). The five-year deal was estimated to have been responsible for as much as one-fifth of THQ's revenues in that period, leading analysts to speculate that the deal's end would plunge THQ into bankruptcy (Martin, 2009). THQ survived and has continued to make games for Pixar, acquired by Disney in 2006, including helping with a 2012 adaptation of the film *Brave* (2012) (Conditt, 2012). In this regard, 2004 marked a turning point in the industry, with a range of game companies seeking to take advantage of animated properties. In addition to THQ's successes with adaptations for *The Incredibles* (USA) and *The SpongeBob Squarepants Movie* (2004), Activision's adaptations of *Shrek 2* (2004) all sold well. In the same year, game publisher Ubisoft agreed to a multi-picture deal to produce games based on two of Sony Pictures' animated movies, *Open Season* (2006) and *Surf's Up* (2007) (Fritz, 2005i). In spite of that fervour, movie tie-ins were not guarantees of success. Of all the video games based on films in 2004, only that based on *Spider-Man 2* (2004) made it into the top ten sellers for that year (Levine, 2005c). Though fortunes have been mixed, Hollywood and the video game industry have continued cementing their ties, particularly focusing on children's games and comic book properties (Brookey, 2010). Among the higher-profile deals was a five-year exclusive arrangement between Nickelodeon and THQ, reportedly worth more than $75 million (Fritz, 2005a).

One of the more interesting licensing trends sees video games delving into Hollywood's extensive back catalogue. Following the success of major franchise licensing deals like *Harry Potter* and *The Lord of the Rings*, games have been based on older movies (Wingfield and Marr, 2005). One example is EA licensing the rights to *The Godfather* (1972–90) trilogy from Paramount (Grover *et al.*, 2005). Similarly, Take-Two Interactive released a game based on the cult hit *The Warriors* (Wingfield and Marr, 2005). Plans for a game based on the Dirty Harry character, in conjunction with Clint Eastwood's Malpaso Productions and Warner Bros. Interactive Entertainment (Guth *et al.*, 2005; Robischon, 2005) were ultimately cancelled. It was reported to include scenes and characters from across the franchise, though no reference to the previous Dirty Harry game, made in 1990 for Nintendo, was expected (Robischon, 2005). Other movies from the vault viewed as likely transferrals include *Taxi Driver* (1976) and *Scarface* (1983), produced by Majesco Entertainment and Vivendi Universal Games respectively (Wingfield and Marr, 2005).

This move to exploit Hollywood's catalogue makes economic sense as the rights to older films come much cheaper than developing a new concept. Estimates suggest such rights can run between US $150,000 to $400,000, much cheaper than the US $1 million or more it might cost for a new release or an entirely new concept. Unlike releases based on current movies, such games lose the benefit of the marketing provided by the promotion of new films. Because major games can now cost more than $10 million to produce, without such marketing, a licensed game may need to sell as many as half a million copies to be profitable (Levine, 2005c).

Perhaps no better example of Hollywood's exuberant adoption of video games can be found than the recent *Enter the Matrix* (2003) online game developed by Shiny Entertainment and published by Atari (Herold, 2003b).With a budget of roughly $20 million, higher than most, the game included an hour of filmed scenes and another hour of animation (Delaney, 2003). The game, released in 2003, more than two years after the last film in the *Matrix* trilogy (1999–2003), was expected to bring in $500 million between 2003 and 2006 if successful. Unfortunately, it has proved relatively unpopular and received bad reviews, resulting in Warner Bros. selling the rights to Sony (Fritz, 2005h).

Such examples suggest that the future may not be all roses for video games and Hollywood. Recent Microsoft plans to work around Hollywood and develop its own movie for the *Halo* franchise have raised eyebrows in the film industry (Fritz, 2004; Grover *et al.*, 2005). Though not a direct response, Hollywood companies are considering tightening policies on video game production. The head of Warner Bros.' game division, Jason Hall, stirred the hornets' nest by insisting that licensed games that do not meet minimum quality standards set by Warner Bros. will receive reduced royalties (Holson, 2005b).

Another area of confrontation centred around the growing demand for Hollywood talent in video games. Games are beginning to require A-list voices to help them succeed (Brodesser, 2005). Hollywood and the Screen Actors Guild have been insisting on higher royalties for voice work, something that the video game industry isn't certain it can afford (Fritz, 2005f, 2005g). In most cases, unless a game company has permission from an actor, it cannot use the performer's likeness (Wingfield and Marr, 2005). This has led to some strange conflicts, including the absence of Michael Corleone (who would have been voiced by Al Pacino) in the games based on *The Godfather* trilogy. Even the action star turned California governor, Arnold Schwarzenegger, has begun to license his voice for games based on *The Terminator* franchise (1984–2003) (AP, 2003). As will be seen in Chapter 5, the video game industry – like most high-tech industries – has little experience in dealing with unions and guilds.

Finally, the increasing sophistication and availability of game design programs included with many video games has begun to encroach on Hollywood. A number of games include built-in movie-making tools allowing players to make and modify their own animation (Economist, 2004). This animation, termed 'machinima', can rival the graphical capabilities used to create movies like *Shrek* (2001) or *Finding Nemo* (Economist, 2004; Levine, 2005b). Machinima has been used by Spike TV to help create shorts for its 2003 video game awards programme and by Steven Spielberg to storyboard his movie *A.I.: Artificial Intelligence* (2001) (Economist, 2004). It has also become a staple of video game marketing campaigns and has influenced the development of a number of television shows and Internet shorts based on video games (Levine, 2005c).

GAMES AND TELEVISION

One crucial industry impacted in various ways by video games is television and cable. Perhaps surprisingly, in view of the often tight relationship between games and film production, particularly in Hollywood, that between television and video games has evolved in an entirely different direction. Like motion pictures, television and cable have been forced to come to grips with the increasing popularity of video games. By 2003 United States' viewership for top broadcasters fell almost 22 per cent in the eighteen to thirty-four age bracket and was down 7 per cent overall (Donaton, 2003). By 2012, this fall had yet to stop, with broadcasters reporting losses of between 3 to 21 per cent of viewers in the eighteen to forty-nine demographic (Carter, 2012).

In spite of these losses, broadcasting and cable TV have been relatively limited in their use of video games to attract audiences. Unlike film, it has seen limited success licensing its content to game makers, while the adaptation of games for the small screen has been primarily restricted to animation. Estimates suggest that in 2002, cable network Nickelodeon's various licensing deals brought in more than US $2.5 billion in sales (Elkin, 2003). Toy and video game producer Bandai signed deals with various networks including WB, the Disney Channel and the Cartoon Network to make games and toys based on their properties (Fritz, 2005e). Fox signed a deal with legendary game maker Will Wright, creator of the *SimCity* (1989) and *Sims* (2000) franchises among others, to create original programming for the network (2003).

Instead, attempts have been made to bring video games into more standard content. In 2003, two series, *Pirate Islands* (2002) and *.hack/sign* (2002) featured characters trapped in video games (Herold, 2003a). While there are certainly other examples of such attempts, it's worth noting that both series only lasted a season. More commonly, networks like Spike TV, MTV and the Game Show Network have become involved by focusing on awards for game design, competitions or shows focused on reviewing games or providing tips on play (Wired, 2003). Spike TV hosts annual game awards, which feature not just games but also rock stars, athletes and other celebrities while also serving to promote upcoming video games (Breznican, 2004). In the US, cable provider Comcast launched its all video game network, G4, in 2002 (Sieberg, 2002). At the time of its launch, the network was estimated to have reached between 15 and 54 million viewers, and this led to its expansion into Canadian cable (Stanley, 2004a). Though it was plagued with concerns that its remit was too narrow, the network continues to run (Kelly, 2006). Because licensing is becoming a vital source of revenue for television and cable, further exploration of video games seems likely.

No network has been more aggressive in developing content and licensing with video games than MTV. The network has been at the forefront of promoting games with its advertising and in its programmes. Typically, it offers promotional time in exchange for a take of game revenue. MTV's first agreement was with Midway Games, then owned by Sumner Redstone. The first game to come from this alliance was based on the television show *Pimp My Ride* (2004) and was released in 2006. As part of the arrangement, MTV consulted on the game's soundtrack, sold in-game advertising and lent its logo to the package (Levine, 2005a). MTV has also experimented with machinima. In addition to the use of video game content and production in videos, the company aired a show, *Video Mods* (2004), on its second network, MTV2. The show featured music videos designed by machinima and showcasing popular game characters (Donaton, 2005). In spite of the show's popularity, MTV has not been able to further capitalise on it in other media because of the rights' questions raised from both its use of music and game characters.

Part of the difference in the relationships with film and television/cable industries is due to the video game industry's own preferences. Owing to the planned obsolescence of hardware as well as the increased capability both in terms of graphics and media, manufacturers in other media industries have come to rely on video games as drivers for new hardware purchases. Television manufacturers are a prime example of this, Sony with perhaps the best, having focused its development of the PlayStation systems as a means of bolstering sales of its televisions and those of its strategic partners like Samsung. In contrast, Microsoft has emphasised that it sees its Xbox platforms as the future of media in the home, and in some cases suggests Xbox is really a competing cable channel (Wilson, 2011).

GAMES AND RECORDED MUSIC

Of all the media industries, recorded music has one of the longest and most consistently profitable relationships with video games, which were beginning to explore ties with music as early as 1982. Atari's *Journey Escape* featured an 8 bit version of the song 'Don't Stop Believin'' by the rock band Journey in a game designed to help promote the band and their album *Escape*, released the same year. As technology improved, it became possible to offer both more sophisticated renditions of music and a larger selection. As Ben Aslinger notes, it only took about fifteen years for video games to move from simple 8 bit synthesised recordings like those in the Journey game to 'popular songs' saturation' of games (Aslinger, 2008). Along the way, the ability of video games to promote artists and the recorded music industry developed as well. A more advanced but similar promotional game to *Journey Escape* was launched in 1989 on the Sega Genesis console and into arcades to promote Michael Jackson and his movie *Moonwalker* (1989). The game, titled *Michael Jackson's Moonwalker*, featured a number of the popstar's hit songs including 'Beat It' and 'Billie Jean' (Cerrati, 2006).

By 1994, the use of music in games had gone beyond promoting a single artist. In that year, EA released *Road Rash*, one of the first games to embrace music tie-ins, featuring licensed music as part of the gameplay (Cuneo, 2004). By the early part of the twenty-first century, game companies were signing formal agreements with record labels. EA teamed up with Def Jam, while Activision partnered with Maverick Records, in deals that produced both games and soundtracks for consumption (Aslinger, 2008). Since then, a number of major games have involved new music and record deals. In October 2002, Epic Records released seven albums to accompany the hit game *Grand Theft Auto: Vice City*, featuring music from a wide variety of artists and genres. Artists included in the game included Ozzy Osbourne, Blondie, Tears for Fears, Toto, Luther Vandross, Michael Jackson, Grandmaster Flash and the Furious Five, Rick James, Machito and his Afro-Cuban Orchestra and Tito Puente (Garity, 2003). Estimates suggest that between 2001 and 2004, EA increased its music licensing more than 270 per cent (Cerrati, 2006). This expanded use of licensed music has led some to call video games the 'new radio', serving as an ideal way to secure exposure for new and upcoming releases to focused demographics (Cuneo, 2004).

The rise of music-oriented games like MTV's *Rock Band* and Activision's *Guitar Hero* and *DJ Hero* (2009) series has only pushed the limits of what music licensing can do. Often referred to as rhythm games, these involve the player using a controller to mimic playing various licensed music pieces, typically (though not always) from established musical acts. Such games, once bestsellers, have begun to see declines in the sales of game units, even as the sale of licenses for songs to be used within them, via digital download, topped more than 60 million worldwide in 2010 (Scelsi, 2010). In the case of the 2009 release *The Beatles: Rock Band*, digital downloads of individual Beatles albums cost approximately US $17 while individual songs cost US $2; typically, a band would see about 30 per cent of that cost, though it's assumed the Beatles did considerably better (Bruno and Peters, 2009).

Michael Cerrati provides a detailed analysis of licensing arrangements between the recorded music and video game industries. In one formulation, game makers pay a lump sum at the outset to the copyright owner, but in other, less common cases, game makers may pay per-unit royalties or a combination of the two. Fees for music licensing are a complex area determined by a number of factors including composition, genre, duration and the amount of music used. Moreover, the licensing of music is a twofold process, requiring both a master use licence, which allows use of the recording, and a synch licence, which allows the use of the recording as part of a synchronised presentation. As of 2006, licence fees were typically between US $2,500 and over US $20,000 per composition. Based on these rates, a composer might expect to receive between US $1,000 and US $1,500 per minute of music used in games that sell well (Cerrati, 2006).

One benefit of the rise of music licensing has been its ability to bolster the sagging sales of the recorded music industry; while sales of CDs have continued to decline, EA has estimated that featuring songs in video game soundtracks could improve an album's sales by as much as 40 per cent (Aslinger, 2010). Even established bands can benefit, as estimates suggest that veteran rock act Aerosmith saw a jump in sales of more than 130 per cent for their song 'Same Old Song and Dance' after it appeared in the game *Guitar Hero: Aerosmith* (2008) (Scelsi, 2010). For less well-known bands, it is thought that being featured in a rhythm game can boost sales by as much as 300 per cent (Graser, 2009). Such a return is noteworthy because estimates suggest that for the average Triple-A game, the budget for music licensing is typically only about 1 to 2 per cent (Aslinger, 2009). Such figures show that video game's use of licensed music matches up well with the rise of digital distribution methods such as Apple's iTunes. Less established acts have also been able to take advantage of video games as a promotional tool, as networked game systems allow artists to upload their own tracks to MTV's Rock Band network. Any tracks approved for distribution are made available for sale in the marketplace, generally for between US $1 to $3, of which artists receive approximately 30 per cent (Scelsi, 2010).

GAMES AND PUBLISHING

One area of the video game industry that hasn't received much attention is the publication of strategy guides. These provide walk-throughs, instructions and hints on how to play games and have become big business as games have grown in complexity. In 2004, the strategy guide industry produced roughly $100 million in sales (Snider, 2004). Prima Games, an imprint of Random House, was estimated to control as much as 90 per cent of the industry in 2004 (Market Square Reporter, 2005c). The company, founded in 1990, has had bestsellers as far back as 1993 when it sold more than 1 million copies of *Myst: The Official Strategy Guide* (2004; Prima Games, 2012). In 2006, estimates suggested that 90 per cent of all guides were sold in physical stores, though Prima was already experimenting with electronic distribution (Market Research.com, 2006). It typically releases between 100 and 125 new guides per year, though it has also begun to sell its back catalogue in eGuide format (Market Research.com, 2004, 2006).

Such guides are created for a variety of genres, including first-person shooters, sports and strategy games. Because of the popularity and complexity of the games they are typically designed for, it is not uncommon for strategy guides to sell more than 1 million copies. Prima's guide for *Halo 2* sold more than 250,000 on its first day at a cost of US $16 each. The growing popularity of eGuides has led to deals with GameStop as well as to the company's production of guides for download on numerous electronic devices, such as consoles; tablets, including iPads; and smart phones, particularly the iPhone (Business Wire, 2012c; Health and Beauty Close-up, 2012).

In addition, more and more 'Making of ...' guides to games are released and becoming bestsellers. These guides may feature tips on gameplay but also focus on stories and artwork created in the game development process. Del Ray Publishing distributed 50,000 copies of its book *The Art of Halo*, retailing at US $21.95 (Market Research.com, 2004; Snider, 2004).

Books have also inspired video games. Ubisoft has developed games drawing on the work of Tom Clancy, who helped create the successful *Splinter Cell* (2002) franchise. The first two games sold more than 6 million units worldwide. Clancy has created two other franchises: *Rainbow Six* and *Ghost Recon*, for which a number of games have been developed. The first *Rainbow Six* game was released in 1998, while the first *Ghost Recon* release was in 2001. These franchises have been profitable enough to spark Hollywood interest. Paramount was slated to make a film based on the *Splinter Cell* franchise, though the movie never made it out of development (McNary, 2004).

GAMES AND SPORTS

Sports, too, have felt the impact of video games. Since 2000, television broadcast ratings for sports programmes have fallen among males between twelve and thirty-four, while general video gameplay has risen to almost equal the amount of time spent watching televised sports (Schiesel, 2005b). No other genre of video games has taken advantage of this shift, maximising economies of scale and the benefits of licensing quite as well as the sports category. In 2004, these games made up 19 per cent of the $6.2 billion in US software sales (Wahl, 2005). Measuring the impact of video games on sports is difficult but, as Sandy Montag, agent to sports commentator John Madden, puts it, 'John Madden's Q rating is in the top ten of all sports and higher than any current football player, and a lot of that is due to the video games' (Schiesel, 2005b).

For a franchise that hit $1 billion in sales by 2004, that statement shouldn't be surprising (Cuneo, 2004). In 2004, *Madden NFL 2005* for the PlayStation 2 ranked third in games sold for the year. The 2003 version ranked first (Wahl, 2005). For the developers of the *Madden* franchise, Electronic Arts, that's just the beginning. EA, which negotiated for exclusive rights to produce NFL games, is estimated to have paid more than US $100 million for them (Nuttall, 2005). Such high fees seem worthwhile. *Madden NFL '13* (2012) sold more than 1.65 million units in the first week of its release, an increase of 8 per cent from its predecessor (Business Wire, 2012d).

Competing against software publishers Sega Sammy Holdings and Take-Two Interactive in the sports genre, EA has scored a number of coups (AP, 2005a). In 2004, EA signed separate deals with cable network ESPN and a number of American sports leagues: the National Football League (NFL), the National Basketball Association (NBA) and the Collegiate Licensing corporation (CLC). The ESPN deal is a record fifteen-year exclusive arrangement and drew ESPN licensing away from competitor Sega (Fritz, 2005a; Wahl, 2005). The deal is part of a growing trend for long-term arrangements rather than one-off licensing pacts. In exchange for the exclusive use of ESPN's logo and images, EA is believed to have paid more than $750 million with some of the money earmarked for marketing, commercials and other promotion (AP, 2005a). Similarly, EA signed a five-year exclusive deal with the NFL. Perhaps most stunning was the six-year deal with the CLC, which grants EA exclusive rights not only to the American college teams represented by the National Collegiate Athletic Association (NCAA) teams, but also to stadiums and schools for college football games. Financial details are not available, but since the CLC doesn't sell its games internationally, it is expected to be less than either the ESPN or NFL deals. In addition, the deal allows the company to produce games for all consoles and platforms, including handheld devices (Gamasutra, 2005a). Like the *Madden* brand, EA's *NCAA Football 2005* (2004) was a major seller, topping more than 1 million copies in 2004 (EA, 2005). Even the National Hockey League (NHL), representing professional hockey in North America, has signed an exclusive deal with EA. Its game *NHL '13* (2012) proved so popular that in only the first week of its release, more than 3.5 million online games were played (Entertainment and Business Newsweekly, 2012).

Athlete portrayals in video games have raised a number of important questions. The licensing of college athletes has not been without problems. As video games have become more realistic, they have also begun to use athletes' likenesses without compensating them (Branch, 2011; Kaburakis *et al.*, 2012). While this is due to current NCAA rules that prohibit student athletes from earning money while participating, signs point to a likely confrontation in this area in the future. In contrast, EA's NHL games gained somewhat dubious praise for finally including a female avatar in their 2012 edition and for then featuring two female hockey players in their roster of available professional likenesses (Journal of Engineering, 2012). This disparity between male and female representation is not unusual in sports video games and suggests an area for possible improvement, and that the industry still doesn't fully grasp the extent of its audiences.

By 2011, the NFL, Major League Baseball (MLB) and the NHL had each reached what amounted to exclusive licensing deals (Good, 2011). In contrast to these groups, the NBA, however, has taken a different tack. NBA commissioner David Stern described the dilemma this way:

> I was on a panel recently where someone asked me what my worst fear was. It was that as video games got so graphically close to perfection and you could create your own players – their hairdos, their shoes – that there might be a battle between seeing games in person or on television and seeing it play out on a video game. (Schiesel, 2005b)

Rather than risk any one video game company becoming a dominant competitor, the NBA has opted to license to five publishers rather than any single one. Its arrangement, lasting between five and six years, was worth $400 million. The five publishers were EA, Take-Two Interactive, Atari, Midway Games and Sony Computer Entertainment (New York Times, 2005b; Nuttall, 2005).

Internationally, similar trends in sports licensing can be seen. In fact, one of the first sports organisations to engage with video game licensing was the International Olympic Committee (IOC). While the IOC hasn't released details about its licensing deals specifically, licensing has been an extremely important revenue source. Between 2009 and 2012, the IOC partnered with 205 organisations earning revenues of US $957 million (IOC, 2012b). Starting with the 1992 games in Barcelona, the first licensed Olympic video game, *Olympic Gold*, was created by developer US Gold (Takiff, 1992). For the 2000 Summer Games, 2002 Winter Games, and 2004 Summer Games, the committee granted the rights to developers ATD and Eidos (IOC, 1999, 2000). Since then, the IOC has experimented with a range of licensing. Starting with the 2008 Beijing games, Sega has held the official licence. Since then, Sega has also developed the official games for the Vancouver and London Olympics, held in 2008 and

2012 respectively. Sega produced two games in conjunction with each Olympics: one a direct simulation, the other using licensed video game characters Sonic the Hedgehog and Mario. The inclusion of the characters is significant as a signal of the cultural significance of video games (IOC, 2008, 2010, 2012a).

Similarly, the Fédération Internationale de Football Association (FIFA) has had great success in recent years with licensed video games. FIFA has also struck a deal with EA, though there are some competitors. Beginning in 1994, FIFA began licensing games across a number of platforms. The most recent iteration – *FIFA '13* (2012) – had versions for all the major consoles as well as for Apples iOS devices (the iPod Touch, iPhone and iPad). Before the game was even launched it experienced considerable success with more than 1.3 million units pre-ordered for consoles, and a 62 per cent increase in downloads from iTunes from 2012, making it the top grossing game in more than twenty countries at the time of its release (Entertainment Newsweekly, 2012). Prior to release, players had access to a demo version, and more than 4.6 million people in 123 countries took advantage (Business Wire, 2012b). When the game launched, more than 350,000 units were sold in North America (Business Wire, 2012a).

GAMES AND ADVERTISING

As with other media industries, the relationship between video games and advertising is in flux, but for different reasons. While the industry has recognised the importance of marketing and licensing, it is still struggling to find a way to take best advantage of the demographic it is pulling together. In 2004, roughly 36 million people in the US played video games. That figure was predicted to double by 2009 (Stanley, 2004a). In fact, by 2012, it had increased to more than 210 million, an increase of almost 600 per cent (Snider, 2012). Those players include a substantial number of eighteen to thirty-four-year old males, one of the most coveted groups in advertising (Stanley, 2004a, 2004b). In 2003, marketers spent a little over US $414.1 million on advertising in video games; in contrast, they spent roughly US $8 billion on television advertising (Stanley, 2004a). This is just a small proportion of the advertising spending in the US. For 2003, while consumers spent almost US $178.4 billion for media, advertisers spent US $175.8 billion on ads. When institutional ad spending – ads aimed not at increasing sales but at promoting an organisation in some fashion – is factored in, the amount increases to $316.8 billion (Donaton, 2005). By 2011, global advertising spending had grown to over US $450 billion (Barton, 2012). It's worth noting that, while video games aren't tracked in that particular study, one of the major categories of casual games would fall under Internet spending, which has experienced the largest growth. As such, the growth of video games as an advertising medium is difficult to quantify but important to consider.

While a lot of money is certainly spent on adverting video games, the more pertinent question is how video games can serve as a new medium for ad placement. As studies show audiences drifting away from other media and finding ways to ignore or bypass more traditional forms of advertising, video games' potential to deliver advertising is likely to become more important. Estimates suggest that US $3.1 billion was spent on in-game advertising in 2010, and that figure is predicted to grow to more than US $7 billion by 2016 (Takahashi, 2011; Tassi, 2011).

In the context of such spending, it is no wonder the industry is working to ensure its products are well received. This helps explain the reliance on franchises and licensed games despite increasing audience pressure to minimise them. As video games become integral in brand building, their placement into media campaigns has changed. Increasingly, the release of games is being delayed till long after the launch of the products they're tied to. For example, both *The Incredibles* and *Shrek 2* games were released months after the films they were based on, but this may have extended the profitability of the property (Diamante, 2005). The video game industry has adapted this practice, using marketing from previous games to boost its own franchises. Sequels, such as those in Microsoft's *Halo* franchise or 2K's *BioShock* (2007) and *BioShock 2* (2010), have become common practice representing a way to decrease risk and development costs (Wahl, 2005). This has resulted in game companies attempting to integrate marketing from the very beginning of the process of developing a game (Miller, 2005).

Industry expenditure on ads is already significant and seems poised to grow. In 2002, in conjunction with the launch of the PlayStation 2, Sony spent $250 million on marketing in North America alone (Cuneo, 2002). Microsoft topped this amount in promoting the Xbox. Globally, the company spent $500 million, with $350 million earmarked for the US (Elkin, 2002d). By 2008, the industry was spending an estimated US $823 million to sell games (McWhertor, 2009). Because video games still rely on a production cycle geared towards Christmas sales, a majority of advertising dollars are spent in the fourth quarter of each year (Hein, 2002).

But the perceptions of the importance of video games and advertising are changing. By 2009, advertising revenue for the industry was predicted to reach $562 million (Gentile, 2005a). As noted, by 2010, it had easily surpassed that, with more than US $3.1 billion in spending (Takahashi, 2011). While the most recent game marketing has relied on the instincts of the advertisers, moves are underway to make marketing more systematic (Waugh, 2005). First, advertisers are recognising the potentials for promotion offered by video games. The marriage of video games and advertising has been along three broad lines:

- Internet games in which ads surround content
- in-game advertising
- custom-published games as advertisements. (Webster and Bulik, 2004)

This combination has proven particularly effective in helping games to attract word-of-mouth publicity and industry buzz (Waugh, 2005). Internet advertising surrounding content is the most basic means of combining ads with video games. This form of advertising works like any web page. In this case, the main content is the game, but the web page also shows ads. As such, ad sales here are no different than for other websites.

In-game Advertising

One of the big questions for the industry has been how to incorporate advertising directly into games. Advertisers spent $34 million on in-game ads in 2004, considerably less than is spent on most other media, but by 2010, that number had grown to $3.1 billion (Gentile, 2005a). Estimates indicate that spending will increase to $7.2 billion by 2016 (Takahashi, 2011). Advertising in games can work in two ways: static and dynamic (Bulik, 2004; Takahashi, 2011). Static product placement puts advertising into the games in specific ways. The classic example is of an arena in a sports game where advertisements are expected as part of the environment. In static advertising, they are hard coded into the environment. This is becoming increasingly common in sports games that incorporate advertising into arena depictions, but it is possible in other games as well. For example, in EA's *The Sims Online* (2002) advertisements abound, with everything from fast food to apparel and even computers represented (Elkin, 2002a). In a subtler manner, Activision has placed Puma products throughout *True Crime: Streets of LA* (2003), not only in the form of static ads, but to the extent that the main character is also decked out in his Puma finest (Bulik, 2004). Sony is also experimenting with in-game advertising. Hoping to hype its MiniDisc Walkman, the company retooled *Pro Skater* (1999) into an online game christened *Sony SkatePark* (2002). Players played an average of 2.8 times during the game's twelve-week run (Elkin, 2002b).

Dynamic advertising promises to be the more interesting of the two styles. In dynamic ads, the products placed throughout the game change periodically. One such example could be seen during the 2008 US presidential campaign; Barack Obama's static ads appeared in EA's racing game *Burnout Paradise* (Itzkoff, 2008). If player consoles were connected to the Internet, billboards along the racecourse would change to show the Obama ads and others. Obviously, such ads require an Internet connection in order to be effective, but with the most recent console generations, this is rarely a problem. Some companies are attempting to measure the effectiveness of such ads; in *Underground 2* (2004) Activision embedded markers in each Jeep image to help count player contact with the placements (Gentile, 2005b).

Though ad revenue is growing, it must be emphasised that it only represents a small percentage of overall revenue. Electronic Arts earned $4 billion in product sales in 2004, but only $10 million of this from advertising (Richtel, 2005d). As video game advertising matures, however, marketers are already planning ways to launch products from games (Bulik, 2004). It is estimated that publishers – and not developers – could eventually earn from $1 to $2 in advertising per game played. The promise of such lucrative revenues has spawned a new advertising group, Massive, which plans to create dynamic ads in PC games. The company has deals with at least ten publishers, and clients such as Coca-Cola, Intel, Paramount Pictures and Universal Music Group. Massive's biggest deal is with Nielsen Ratings, aiming to determine whether in-game advertisements are effective, using a combination of in-game metrics and consumer polls.

Custom Games

Major companies in particular have begun to experiment with custom-published games, designed specifically to promote a product or brand. In fact, these represent an early example of the 'persuasive games', which develop out of the idea that games have unique rhetorical capabilities through procedural logic (Bogost, 2007). When used for skills training – as in military or law enforcement simulations – they are frequently referred to as 'serious games'. While the main focus of the industry has been on custom games for advertising, such games can also aid journalists and have contributed to the notion of 'gamification', which is essentially the attempt to use the procedural logic and rhetorical ability of games and game features in non-gaming contexts (Bogost *et al.*, 2010; McGonigal, 2011). Examples of this might include virtual badges, to be shown on a social network whenever one goes to a particular shop or video game mechanics helping to solve scientific problems, as in the online game and website *FoldIt*.

While the possibilities are broad, advertising seems to have taken much of the focus of custom games. Companies including Jeep, Nike, Volvo, Levi Strauss, Coca-Cola, Nokia and Kraft Foods are experimenting with custom games to promote products (Bulik, 2004; Stanley, 2004a). Even the US Army has developed custom games to help them meet recruiting challenges (Huntemann and Payne, 2009; Nichols, 2009). These games draw on the casual gaming trend and allow advertisers a great degree of control over how their brand is presented

(Diamante, 2005). The design simplicity of casual games makes them extremely flexible for international marketing. The game mechanics are usually fairly straightforward, meaning that they work across a range of cultures with only small amounts of accompanying text needing adaptation. One game designed by WildTangent for Nike, called *Skorpion K.O.* (2002), was released in eleven different languages and played by more than 600,000 people worldwide (Elkin, 2002b). Another Nike-sponsored game, created in conjunction with Weiden and Kennedy advertisers titled *Game Breakers* (2003) was so popular that its makers considered expanding it into a fully fledged game (Stanley, 2004a). How effective are these games? According to Chrysler, which put a variety of simple puzzle and sports games on its website as well as on CDs distributed in magazines, 3.5 million people registered on the site and downloaded games. Of those, roughly 10,000 eventually bought Chrysler vehicles (Gentile, 2005b).

CONCLUSION

One of the major challenges for the video game industry has been the need to identify new revenue streams to help insulate itself from risk. Rather than function in an insular fashion, the industry has sought out ties via licensing and integration with other cultural industries. Currently, the longest running and most impressive ties are with film and organised sports. Both allow for easy licensing of products that maximise the revenue to be gained from a concept. This, no doubt, is owing to the industry's similarity to Hollywood in terms of organisation, discussed in Chapter 1. While licensing of products between film and video games has been undertaken from both directions, the current power dynamic seems to favour Hollywood, if only because of its considerably larger catalogue of titles, which the game industry has started to tap for major products. Sports licensing offers similar benefits, allowing video games to draw on the regular changes in team rosters in order to make continued licensing and updates necessary. At the same time, sports licensing also offers a second potential revenue stream through in-game advertising.

The high levels of concentration discussed in Chapter 2 also influence how the industry relates to other media. This is particularly true in the area of licensing. By consolidating their power, game publishers have ensured their ability to draw more revenue from licensing and advertising than any other segment of the industry. As seen with Electronic Arts' signing of exclusive licensing deals across a range of sports, dominance in a particular genre of games becomes hard to disrupt. Sports video games not only serve as a consistent revenue source but also represent highly contested battles for licensing rights. At the same time, licensing to video games presents new challenges as well. The example of the NCAA's licensing of student athlete images as well as battles over music rights suggest that video games are likely to present new challenges to licensing contracts in coming years.

The incorporation of advertising into games is perhaps one of the most important developments. Seen as a growing area, several companies have been seeking ways to make their products useful to business, education and government sources. In part this owes to the links between the industry and advertising. Advertising spent on games is significant, but the revenues drawn from selling ads are growing rapidly. While the marriage of games and advertising allows for obvious revenue benefits, it has also opened up new segments for game development, such as custom video games. These games have, in turn, prompted new genres, like the persuasive or serious game as well as gamification, the idea of using game mechanics and rules in other fields.

With an understanding of the industry's structure discussed in Chapters 2 and 3 and its ties to other media industries, assessed in this chapter, it is possible to focus on the production of the commodities themselves – the hardware and the software which make up the chief products – as well as on labour conditions and the educational requirements for employees in the next chapter.

5

Labour and Production in the Global Video Game Industry

The historic development of the video game industry has resulted in a very particular set of logics of production. First, it is a business that produces commodities geared towards planned obsolescence roughly every two to three years. Second, its production schedule revolves around the Christmas buying season and opportunities to take advantage of the marketing of related licensed products, in particular films and television shows. Lastly, the industry has sought to expand its market both in terms of who plays video games and where and how they are played.

With this in mind, this chapter focuses on the production of the commodities themselves. In order to better understand the production process, particular attention is paid to the situation of workers, as well as how they are organised and educated. It must be remembered that there are two key areas of production: hardware and software. The hardware side tends to be contracted out following established patterns of chip manufacture. The second level is that of production of the games themselves. This tends to be where the most information is available and serves as the focus of most studies. The business ties to other media industries, particularly film, have resulted in tensions about how labour is organised that distinguish video games from the rest of the software industry.

On the whole, the video games industry is an ideal example of what it means to live and work in an information society. Programming and design jobs require both high skill levels and, in many cases, high levels of creativity. Thus workers tend to be highly educated. In addition, the reliance on computers and high-speed Internet allows positions on the software side to be highly mobile. This combination of skills and education on the part of employees is one of the major value-added benefits of game software, in particular. Creativity is valued by the creators and consumers of games alike. As such, software workers are excellent examples of creative labour. Owing to a rare combination of skills that are in great demand, video game workers should be more able to switch jobs within the industry and more secure in their jobs. It would be expected that they would earn more money, have better benefits and experience greater job satisfaction than employees in most other sectors. This, in turn, should result in higher productivity compared to other non-information industries. As this chapter shows, however, there are limits to these benefits for workers.

This chapter examines the similarities between the video game industry and the computer and information industries, and lays out the challenges and constraints in producing video games. It debates the realities of employment in the hardware and software sectors. It looks at wages, education and challenges around unionisation and gender representation. It also explores the growing importance of education programmes and discusses ways that game players have become important factors in the production process.

LABOUR IN THE COMPUTER AND INFORMATION INDUSTRIES

In many ways, patterns in video game labour resemble those in computer production more closely than they do other creative industries. This is true even of film, despite the similarities demonstrated in previous chapters. By the start of the 1970s, labour patterns in high-tech manufacturing had split along two lines: highly sought after human capital and innovation on one hand, and cheap manufacturing capability on the other. On the software side, the reliance on human capital – the investment in training – has become a driving force, while on the hardware side globalised outsourcing and supply-chain management has become the order of the day (Carillo and Zazzaro, 2000; Nichols, 2013). That these features exist within a perpetual cycle of planned obsolescence makes the video game commodity work particularly well in the globalised, information economy because there is a means of influencing demand which helps to drive profits while allowing producers to take advantage of differing labour situations (Dyer-Witheford and De Peuter, 2009; Kline *et al.*, 2003). Only some of this can be attributed to technological shifts and economic upswing even in the information industries. These industries, including the high-technology and computer sectors, have been in a state of growth for over a decade. In the year 2000, one study found that, while more than 1.6 million technology jobs were created each year, almost half would go unfilled (Obermayer, 2000). The range of creative industries policy initiatives around the world since suggest that jobs in video games are seen as worthwhile and in demand. Incentives programmes including tax breaks and other government subsidies to help develop production have been on the agenda around the world, including the US, UK, Canada and Australia (Comas, 2012; Sherwin, 2012; Vara and Woo, 2012).

In the Internet sector companies such as AOL have created more than 100,000 new jobs between 1995 and 1997. Despite this proliferation of posts in computing, there are few examples of labour unionisation; workers are instead lured with promises of stock options and public offerings (Ross, 1999). The service sector in the US accounts for more than 75 per cent of the Gross Domestic Product (GDP). The high-tech and dot-com industries have been a substantial part of this (McCammon and Griffin, 2000). However, in 1998, this wild growth came to a dramatic halt, ultimately leading to layoffs (Race, 2001). These layoffs have spelled difficulty for the industry.

The software publishing sector, which includes video games, is ranked by the US Department of Labour as the fastest growing industry in the US economy. In fact, it represents only a small proportion of overall employment in the US. According to the US Bureau of Labour Statistics, in 2005 128,827,360 individuals were employed in the US. However, less than 1 per cent of the total population is involved in the game industry and their wages are largely comparable to other sectors of the economy (BLS, 2005a, 2005b). By 2010, more precise measures indicated that more than 32,000 people were directly employed by the US publishing sector of the industry, with approximately 120,000 people employed directly and indirectly in the US (Siwek, 2010). Similar numbers were seen in the UK, where one report estimated that more than 7,000 people were directly employed by more than 155 game development companies, with estimates suggesting that more than 39,000 more employees could be found in related software development industries (Kerr, 2012). The size of the industry in these two areas reflects their importance in the global market and is impressive when compared with figures in countries where game development is still growing. One example is Sweden, where by 2009, the industry was estimated to employ almost 1,600 employees in total (Sandqvist, 2012).

Moreover, as with video games, the software industry as a whole is consolidating, with less than 7 per cent of employers accounting for more than two-thirds of jobs (DOL, 2005). Employees are typically younger than workers in other industries (Zito, 2000). In addition, as jobs begin to migrate overseas, the perception of employment in the software industry is becoming more and more negative (Konrad, 2005).

One of the key factors distinguishing game employees from other creatives centres around labour organising and unionisation. Unlike the highly unionised worlds of film and music production, which exhibit patterns in terms of skills and education required, video games instead mirror labour patterns in computing (Dyer-Witheford, 2002; Wasko, 1998). As noted, in the computer industry, despite the high demand for jobs and increasing globalisation, there is little unionisation. This pattern has been true for video games as well (Goetz, 2012; Lacina, 2012; Shayndi, 2012). Like other high-tech industries, because the video game business has been, until recently, slow to offer information about its labour situation, getting a sense of unionisation has been difficult. For high-tech manufacturing in general, however, only a small fraction of workers have unionised (Pfleger, 2001). As worker dissatisfaction grows, something discussed further in this chapter, there is an increased likelihood that workers will unionise.

In the wake of the global financial crisis, video games went through a round of company closures and layoffs (Denison, 2011; Gaar, 2011; Lee, 2012; Sherr, 2012) sparking new worries for workers. Closures resulted in submerged value of stock options offered to employees – one of the standard benefits of high-tech companies (Race, 2001). Like the dot-com bust before, these changes forced workers to reconsider the value of job security, eight-hour work days and working in a meritocracy (Smith, 2000). Moreover, there is a second, often ignored portion of the high-tech labour force. Corcoran points out the value of thousands of volunteers and legions of part-time and temporary workers who have made the high-tech economy possible. Many have spent considerable time and effort organising chat rooms, online games and contributing programming to the full spectrum of companies in this sector (Corcoran, 2001). The industry also relies on a core group of volunteers called 'Beta Testers' to help in the production of games, a fact that has received little documentation or examination.

Perhaps most interesting, a number of distinct trends in employment can be seen across the information sector, the larger part of the economy video game and software development fall into. The modern economy has gravitated towards the production of knowledge; this shift has been marked by a change in labour patterns and related demands for new skills and the creation of new products (Machlup, 1962; Porat, 1977). Both shifts rely on a move from industrial production to information and service industries. That these industries emerge only in conjunction with telecommunications and computers – a convergence termed 'telematics' – is a fact that is now tacitly assumed. This should not be taken to suggest that the rise of telecommunications and computers paved the road for the service sector. In fact, service industries have existed in some small form for at least as long as industrial production. Teachers and accountants, for example, are both examples of service sector labourers. Instead, these new technologies enabled these sectors to come to dominance within the US economy.

However long-standing service sector employment might be, for the majority of the twentieth century, it has been dominated by women (Bell, 1973). Jobs within this sector have tended to pay low wages, be poorly and sparsely unionised, and – contrary to much of the rhetoric about information economies – low skilled (Martin, 2002). It is becoming increasingly clear that information and service industries have lost considerable ground to global competition. As such, national and regional markets must be understood not just in terms of localised patterns of consumption, labour and regulation, but must also be seen as responding to globalised forces. However, video game labour and production must also be understood in a global context, particularly as supply chains and outsourced labour have become more common.

Globally, a growing tendency towards working longer hours for lower wages can be seen in even the richest nations (Miller, 2008; Nichols, 2013; Yates, 2001, 2003). Moreover, because there is an increasing trend towards a 'horizontal labour market' in which people shift jobs frequently within an industry rather than moving up within it, the prospects for employees in the business may be bleaker than they are often portrayed (Florida, 2002).

The high-tech industries have also evidenced an alarming trend towards gendered labour practices, with female employees excluded from the best jobs and receiving consistently lower wages for the few remaining positions (Martin, 2002). This is particularly troubling as the industries involved globalise production. In the video game industry, where much of the content has been historically directed towards males, it would not be surprising to see these trends continue.

LABOUR AND THE PRODUCTION OF VIDEO GAMES

Labour in the video game business functions differently in the hardware and software sectors. As noted, labour on the software side is of the high-tech and creative variety and has received considerable attention. In contrast, the production of the hardware ranges from industrial mass production at its best to near slave labour at its worst. Perhaps unsurprisingly, these two sectors are largely separate, with most software development happening in the industrialised countries, also the major consumers of the products, while hardware manufacture takes place in the global marketplace wherever the labour is cheapest and working conditions least restrictive (Miller, 2008; Nichols, 2013).

North America, Western Europe and Japan make up a majority of video game sales, whether hardware or software. However, most of the hardware production of items like microchips and semiconductors takes place in various Asian countries, including India, China and Taiwan (Burt, 2005; Mazurek, 2003). To take advantage of centralised production as well as the cheaper labour and transportation costs, the three major hardware manufacturers have all set up most of their production in the same region of China (Daily Gleaner, 2005; Takahashi, 2005). Microsoft perhaps best exemplifies this strategy. As discussed previously, the production of its first console, the Xbox, was estimated to have lost the company as much as US $125 per console, and so Microsoft worked to shrink its supply chain when it began to produce the Xbox 360. Manufacture of the Xbox 360 includes more than 1,700 parts from 250 suppliers and is estimated to involve more than 25,000 workers worldwide (Guth, 2005).

MINERAL AND RESOURCE PROCUREMENT
MINERALS GATHERED FROM A NUMBER OF COUNTRIES
AND PROCURED BY UNAFFILIATED CONTRACTORS

RESOURCE TRANSPORTATION
CONTRACTORS ARRANGE TRANSPORTATION OF RESOURCES
TO KEY ASSEMBLY LOCATIONS. THIS IS THE FIRST CONNECTION
TO THE MAINSTREAM INDUSTRY

COMPONENT MANUFACTURE AND ASSEMBLY
COMPONENTS ARE ASSEMBLED INTO CONSOLES AT
INDUSTRY-OWNED OR DIRECTLY OUTSOURCED FACTORIES

DISTRIBUTION TO KEY MARKETS
CONSOLES ARE TRANSPORTED TO DISTRIBUTION CENTRES
WHERE THEY CAN BE SENT TO NATIONAL MARKETS FOR SALE

Figure 5.1 Hardware Resource Supply Chain and Production Process

The cheaper labour costs are one of the primary benefits of the globalised manufacture of game consoles. This is true throughout the supply chain process,
described in Figure 5.1. Hardware manufacture, unsurprisingly, is much more
industrialised than information or service provision in its nature, with many of
the tasks relying on the most manual of labour. Resources for the creation of consoles and other devices are typically extracted and then shipped to central sites
for processing and assembly. The physical resources come disproportionately
from developing countries. As noted, China is the most likely point of manufacture. Once the hardware is put together, it can then be distributed to the typically
heavily industrialised markets for video game products. As will be seen, the global
supply chain for console manufacture involves a number of countries, only a few
of which are heavy game consumers. In addition, many of the raw materials for
console production are not procured directly by the manufacturers but by intermediaries. Estimates suggest that workers in the Chinese factories engaged in
manufacturing consoles typically earn around US $200 per month, much of
which is paid back to the factories which employ them for room and board as

well as for any failures to meet their assigned production goals (Ho *et al.*, 2009). Unlike in software production, these workers are predominantly female migrants, sending money back to families in rural areas (Pöyhönen and Simola, 2007).

A second benefit is that the globalised supply process allows the major manufacturers to distance themselves from alleged abuses in parts of the supply chain, particularly those associated with the procurement of the raw materials. For example, when Sony brought its sixth-generation console, the PlayStation 2, to market in the year 2000, it experienced considerable delays due to problems getting the rare earth mineral Coltan, found primarily in the war-torn Democratic Republic of Congo (DRC) (Vick, 2001). By 2011, China had become the world's leading supplier of this and a number of other refined minerals for high-tech devices, controlling as much as 97 per cent of the world's production (Forbes, 2011). The fact that these high-tech devices rely on this mineral led some to describe the conflict in the DRC as 'the PlayStation War' (Game Politics.com, 2008; Lasker, 2008). While Sony has denied using Coltan from the DRC, reports estimate that, with roughly 80 per cent of the world's Coltan originating there, it would be very difficult to avoid (Dyer-Witheford and De Peuter, 2009; Vick, 2001). Table 5.1 details some of the minerals, particularly conflict minerals, used in game console production and the countries they come from. Reliance on intermediaries to procure the minerals and other raw materials has meant the major manufacturers have not been able to ensure that they aren't using conflict minerals. This has led to criticism from environmental organisations, particularly Greenpeace, for lagging behind even PC manufacturers in making their products more environmentally friendly (Greenpeace, 2007a, 2007b, 2008; Orland, 2010). The reliance on periodic obsolescence has also resulted in problems in the form of the discarded consoles. Because so few are recycled, the environmental challenge posed by consoles extends beyond the question of production (Maxwell and Miller, 2012). This outsourcing is part of a broad trend in the industry, which has ramped up its use of the practice considerably between 1995 and 2006; the Asia-Pacific region was estimated to have increased its percentage of total global high-tech production – including games consoles – during this period from approximately 20 per cent to more than 40 per cent (Byster and Smith, 2006; Ho *et al.*, 2009). Another benefit of moving work to this region has been the lack of unionisation across most of the production chain. Even where unions are in place, there can be problems. For example, unions in China are run by the government and have been seen as lax because reporting union difficulties would mark a contradiction and a failure in the country's socialist philosophy (Nichols, 2013).

While the manufacture of consoles and the creation of microchips can be particularly hazardous because of the toxic materials that are used and emitted, which can have considerable impact on both employees and the environment,

Country	Resource
Australia	Coltan
	Tantalum
Bolivia	Tin
Brazil	Tantalum
	Tin
China	Beryllium
	Gallium
	Tin
Democratic Republic of Congo	Cobalt
	Coltan
	Tantalum
	Tin
Ethiopia	Tantalum
Indonesia	Tin
Mozambique	Tantalum
Peru	Tin
Russian Federation	Cobalt
	Palladium
Rwanda	Tantalum
South America	Platinum
	Ruthenium
Ukraine	Gallium
Zimbabwe	Ruthenium

Table 5.1 Key Mineral Resource Production by Country
Sources: Guth (2005); Kim (2011); Nichols (2013); Takahashi (2005).

just as dangerous is the extraction of the raw materials themselves (Kuehr *et al.*, 2003; Williams, 2003b). Semiconductor production carries with it the danger of both cancer and reproductive disorders, particularly troubling since the labour force is predominantly female (LaDou, 2006). Incidents at Apple's FoxConn plants, including employee suicides and exposure to dangerous chemicals has focused attention across the industry (Barboza, 2011; Luk, 2011).

In contrast, the creation of game software has situated itself in the primary markets for game sales. As has been noted previously, the increasing technical capabilities of hardware platforms, particularly consoles, help to drive the production of new video games. This has also resulted in higher production costs and longer production times. Since the industry is constrained by the two goals of meeting Christmas demand and promoting brand recognition and licensing obligations, there is intense pressure on workers.

Game production may take between fourteen months and three years (Levine, 2005c; Wattenburger, 2005). For Triple-A games, the industry equivalent of a blockbuster film, costs routinely top $10 million for the most recent round of hardware platforms like the Xbox 360 and PlayStation 3 (Guth *et al.*, 2005; Richtel, 2005d). At that time, some analysts predicted development costs would rise to between US $15 and $20 million (Gentile, 2005b; Grover *et al.*, 2005). In fact, they surpassed those estimates. By 2010, one study indicated that the average budget for game development was US $23 million (Tito, 2010). Higher figures aren't unheard of. One estimate suggests that the cost to produce and market Microsofts's *Halo 3* was over $60 million (Johnson, 2007). The development of new distribution platforms may cut into this and perhaps help to account for the rise in casual game production discussed previously.

Table 5.2 details the rough process of designing video games and how workers fit in. As the figure suggests, since much of the software development is

	Description	Approximate time frame	Number of employees
Design	Creation of the concept. The concept may be generated internally or brought in via an externally licensed product or other idea.	~ 3 months	Smallest staff, often between 3 and 15 people.
Pre-production	Initial ideas and prototypes for game will appear and are storyboarded and put together so that tasks can be given to specific development teams (for example, music, textures, scenery, etc.).	~3 months	Staff size increases to between roughly 15–25.
Production	Creation of games based on storyboards and prototypes. 'Crunch time' begins, as does increased play-testing.	~ 9 months	Staff size maximises from around 25 to more than 200, depending on the game and the company.
Publication	Product is localised, tested and prepared for shipping. 'Crunch time' extends into this period and may last for up to 3–4 months total.	~ 3 months	Varies

Table 5.2 The Process of Video Game Software Development
Sources: Kerr (2006b); Wattenburger (2005).

	Canada		Europe		United States	
	2009	2011	2009	2011	2009	2011
Programmer	67,937	74,970	46,198	46,801	80,320	84,124
Arts and animation	59,400	66,651	38,152	35,887	71,071	63,214
Game design	61,520	60,240	42,423	38,281	69,266	62,104
Production	87,130	71,500	52,125	56,346	75,082	67,265
Audio	61,250	67,955	40,833	25,500	75,082	65,658
Quality assurance	39,375	43,125	29,500	32,500	37,905	45,081
Business	58,929	100,938	59,231	47,222	96,408	79,269

Table 5.3 Changes in Average Software Development Salaries in the US, Canada and
Europe, 2009, 2011 in US Dollars

Sources: Miller (2012); Sheffield and Fleming (2010).

built around deadlines, scheduling tends to work from the deadline backwards, resulting in a 'crunch time' period, with employees typically required to work extended hours in order to make the deadline. The impact of crunch time will be discussed in more detail in the case study later in the chapter. It is also worth noting that, depending on the size of the development studio, the number of employees working on a game can vary. This suggests that in crunch time periods, workers may need to be brought in from outside. This is often the case of play-testers, particularly during the testing and publication phases.

Globally, the primary centres of game software production are North America, Japan and Europe (Nichols, 2013). Table 5.3 compares the salaries across the US, Canada and Europe between 2009 and 2011. The table outlines seven main areas of work associated with the software side of the industry: arts and animation, programming, game design, production, audio, quality assurance and business. Employees in the arts and animation division focus on developing the visual look of things within the game – textures, animations, movement, interface – while others might work on the physical appearance of the product (Duffy, 2007a). Game designers are responsible for guiding the structural elements of the game. They may focus on the overall plot or story, dialogue, level design and flow or scripting. Programmers set up the game's code: the rules and algorithms that make it work and the connections between other types of design such as audio and animation; typically, they account for most of the staff producing a game (Duffy, 2007c). Producers oversee the project and coordinate between the various groups (Duffy, 2007b). Audio workers are responsible for the sound and music within a game, while those in quality assurance test the game and ensure products work across platforms. Finally, the business category includes legal services, public relations, marketing and promotion, accountants and the other staff needed to make game companies run. Table 5.3 indicates some volatility with forward growth in the industry: most sectors have seen earnings grow. But it also shows a clear pattern of higher pay in the North American market, though Canadian workers lead in a few areas.

Manufacturing the physical software commodity is comparable in price to manufacturing a DVD or CD. This means that the majority of the cost of game development is labour, which increases as the process becomes more complicated. In 2000, the US industry paid approximately $7.2 billion in wages to more than 220,000 people in the industry, more than any other country (Aoyama and Izushi, 2004). In the same year, US salaries ranged from around $20,000 to slightly over $100,000 (Deutsch, 2002; Zito, 2000). Entry level game designers made as much as $45,000, while experienced designers could earn up to $120,000 (Deutsch, 2002). At the lower end of the spectrum, paid beta testers typically earned $9–11 per hour and up to $25,000 per year (Hutaff, 1996; Zito, 2000). The work that testers perform, however, accounts for only approximately $50,000 to $100,000 out of a game's production budget. For Triple-A games, this amounts to only 10 per cent or less of the budget. Workloads become heavier, particularly in the polishing phase, with employees routinely asked to work eighty hours or more. EA typically employs forty to fifty testers, but during the summer crunch time that is necessary to release games for the holiday season, it has been known to employ as many as 250. However, smaller publishers and developers often cannot afford their own paid testers and either have to farm the job out or rely on volunteers (Zito, 2000). By 2010, salaries had risen both for software development and within the industry in general. In the US, the average salary had risen to approximately $81,000, while for programmers the average was higher still at $92,000. Audio workers and producers also saw significant increases, reaching approximately $83,000 and $85,000 respectively. In contrast, game designers, including writers and creative directors, as well as artists and animators saw smaller rises, earning approximately $75,000 (Gamasutra, 2012).

For most of the video game industry's history, software labour has looked more like labour in computing than film; it is more Silicon Valley than Hollywood. As power has concentrated, the industry has increasingly relied on stock options and intangible benefits to motivate employees, following the pattern in the computer industry (Richtel, 2005b). Also like Silicon Valley and computing labour in general, there is little union presence. One estimate suggests that fewer than 15 per cent of all games are produced under any form of union contract, though information on which unions is not available (Gentile, 2005a). These concerns may be mitigated in countries other than the US where labour, compensation, vacation time, etc. operate under different social and regulatory norms.

At the same time, the global game business is experiencing the same problems as the larger computer industry. Increased labour costs, combined with labour pressure for better benefits and wages, have resulted in both industries looking abroad for manufacturers. In 2005, Howard Stringer of Sony oversaw a corporate plan referred to as 'Project USA' which, in order to save the company $700 million, cut 9,000 jobs (Reuters, 2005b). The software sector of computing lost 16 per cent of the workforce from March 2001 to March 2004. In the first quarter of 2005, information technology firms in the US laid off approximately 7,000 workers (Konrad, 2005). The global economic downturn hit the industry heavily, and many companies lost jobs in 2008 and 2009 in spite of the industry's overall profitability (Shirinian, 2012). It is possible, however, that those positions are only being relocated to cheaper markets around the globe.

Exporting software jobs to less expensive labour markets has become increasingly important. While India, Korea and China have all benefited from this, Eastern Europe has also become a target, particularly for game development companies in the UK and France (Stone et al., 2006). The value of all outsourced jobs surpassed $16 billion in 2004, with more than 500,000 jobs placed in Bangalore, India alone (Reuters, 2005a). The average programmer in India earns the equivalent of $20 an hour in wages and benefits compared to $65 per hour for US workers (Konrad, 2005). It's also important to realise that the products in question are, like films, increasingly global in nature. As discussed in Chapters 3 and 4, the international markets for video games have come to serve as predictors of what may be successful in the US. The drive to create games that will work in all markets has resulted in a need to localise products. In some cases, this can involve removing questionable content inappropriate for a particular market, while in other cases it may call for additional explanation or content. The continued threat of losing jobs to overseas workers may prompt employees to unionise, particularly in video games, a business industry traditionally resistant to unions (Richtel, 2005c, 2005e). While employees typically receive bonuses at a project's completion, workers in the production, business and quality assurance sides have tended to fare the best, though quality assurance employees typically make the least so bonuses are not as significant. Similarly, as the global economy has struggled, retirement benefits have been reduced (Shirinian, 2012).

EMPLOYMENT IN THE VIDEO GAME INDUSTRY

With governments nationally and locally subsidising video game industries while, as will be discussed in the next section, game design programmes are being created at universities around the world, it is clear that work in the industry is seen

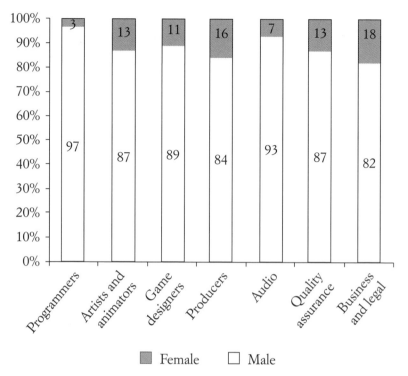

Figure 5.2 Gender Breakdown by Job Category Internationally, 2011
Source: Miller (2012).

as desirable. For this reason, more attention is being paid to understanding who is working in the industry. Perhaps the most important trend to note is the disparity in gender.

Information technology (IT) has demonstrably higher levels of male employment (Martin, 2002). Video games are no different, with a 2004 study showing 92.9 per cent of employees to be male (IGDA, 2004b). Female employment in the sector was higher in the UK during this time period but not by much, making up almost 10 per cent of the workforce (ELSPA, 2004). Since then, the numbers have improved, though they are still of concern. Figure 5.2 displays the gender breakdown by job sector for 2011. It shows improvement in a number of sectors, but still indicates a high level of disparity by gender. Across every major job sector within the industry, women represent a very small minority. Not surprisingly, this has led to criticism in no small part because of the growing importance of female players, seen in a number of markets and noticed by policy makers. Employees are also typically much younger than in other industries. In 2004, approximately 18 per cent of employees were over the age of thirty-five; the remainder were between eighteen and thirty-four in age. In keeping with this, most employees reported that they've been working in video games for less than eight years, putting them at the lower end of the wage scale (IGDA, 2004b).

Job satisfaction has been one of the key issues faced by those in game design. According to a 2004 study by the UK's Entertainment and Leisure Software Publishers Association (ELSPA), the typical worker stays in the industry only three years (ELSPA, 2004). This trend is seen in the US as well. Most US employees indicated that game production wasn't their only choice of career; in fact, 34 per cent planned to leave within five years (IGDA, 2004b). Surveys demonstrate that two factors have more impact on salary in the sector: years of experience and gender (Olsen, 2001, 2002, 2003). Table 5.4 compares salaries by job type and years of experience. Wage disparities by gender abound, with female employees likely to make less than their male counterparts with the same experience in all career paths. In 2002, the only area where women out-averaged men was quality assurance (Olsen, 2004). Table 5.5 breaks down comparison in wages by gender for the range of career sectors. In every sector, women make considerably less than men, with the most sizeable differences – 17 per cent or more – happening in the fields of arts and animation, programming, and business and legal. The disparity in the business and legal sector is particularly problematic because it is one of the sectors with the highest number of women workers. Similar problems have been reported in Silicon Valley in relation to race and ethnicity, though there is minimal data about the video game industry specifically (Greenfield, 2013).

Not surprisingly, the highest wages go to management rather than to creative positions (DOL, 2005). Exact figures are hard to obtain, not just because of the transnational system of labour and ownership, but also because the systems of categorisation of work in the US were changed to a new form (DOC, 1997). Moreover, the measurement scale is not precise enough to separate video game

	Less than 3 years experience	3 to 6 years experience	More than 6 years experience
Artists and animators	49,481	63,214	97,833
Game designers	50,375	62,104	89,231
Producers	55,893	103,080	67,265
Programmers	66,116	84,124	113,694
Audio	32,500	65,658	108,690
Quality assurance	37,500	45,081	61,029
Business and legal	71,818	78,264	123,864

Table 5.4 Average Wages by Job Sector and Years of Experience in US Dollars, 2011
Source: Miller (2012).

	Males	Females
Artists and animators	79,124	52,875
Game designers	74,180	62,000
Game programmers	93,263	83,333
Producers	87,119	78,354
Audio	83,963	72,500
Quality assurance	49,196	39,375
Business and legal	108,402	73,534

Table 5.5 Average Salaries by Gender and Job Sector, 2011 in US Dollars
Sources: Miller (2012); Sousa (2010).

labour from other information technologies in the economic census. What does become clear is that the industry is moving more towards international labour and international audiences (Dyer-Witheford, 2002), primarily to cheaper available labour via globalised production. The production of game commodities in those markets is also likely to heighten demand, even as the industry will need to expand beyond the markets it has focused on if it wants to increase its consumer base.

Even within particular nations, there can be considerable variance in wages by region. The clearest example of this is in the US. The Entertainment Software Foundation (ESA), which monitors the North American market, divides the US video game industry into four main regions. Table 5.6 provides a breakdown of the major regions of production and their corresponding average salaries. The differences seem to reflect a reliance on the computer industry's organisational structure. The highest paying jobs are centred on the West Coast, which holds the Silicon Valley area as well as major software development in the Pacific Northwest (Olsen, 2004) The East region has moved into a clear second place, while the Midwest and South regions have fallen behind. This is probably due to the organisation of the high-tech industries around tech centres, particularly major cities in the Northeast (Mosco, 1999).

Not surprisingly, the video game business's similarities to the computer industry and its interconnections with other media industries have also resulted

	2002	2011
East	62,267	74,796
Midwest	61,970	68,114
South	62,447	69,984
West	79,932	87,809

Table 5.6 Average Wages for US Employees by Region, 2002, 2011 in US Dollars
Sources: Miller (2012); Olsen (2004).

in some peculiar labour challenges. The following case studies examine these two issues.

CASE STUDY: ELECTRONIC ARTS, ROCKSTAR AND LABOUR

Perhaps the most extreme labour difficulty revolves around the practice of 'crunch time' labour exploitation. Notable examples are two major game publishers: Electronic Arts and Rockstar Games. As discussed in Chapters 1 and 2, EA modelled itself on the Hollywood film industry.

However, unlike Hollywood, EA and the video game industry have evolved with little union representation. This is because the video game industry, like the computer industry, has tended to reward employees with stock options and bonuses based on company performances. Such bonuses have become increasingly rare even as video games have become more profitable. Moreover, because Electronic Arts is the largest software publisher in the industry, exhibiting considerable power, its employees serve as an excellent gauge of industry trends.

Between 2000 and 2005, high-profile labour disputes began to crop up. The first and most important of this period involved software giant Electronic Arts. EA, based in Redwood, California, has become the banner case testing whether Silicon Valley practices will continue to hold sway following the dot-com fall (Richtel, 2005b). The company, which employed more than 5,800 people at the time of the dispute, relied on the industry standard production time period, complete with the mandatory 'crunch time' at the end of the production cycle (EA, 2005; Fritz, 2005b). More than half of the employees, however, worked outside the US, including 1,700 in Vancouver, Canada, which could have complicated matters (Richtel, 2005b). Because the industry is deadline driven, projects which fall off schedule result in workers, who are typically contractual, having to put in extra hours in order to meet the deadline. This is referred to as 'crunch time', and it typically falls at the end of the production and the beginning of the promotion phase of the production cycle. During this period, employees are frequently required to work between seventy to ninety hours per week until the project is finished.

The crisis at EA began when the wife of an employee posted an anonymous blog describing the labour conditions of her husband, an unnamed software engineer at EA. Her complaint indicated that her husband, who earned somewhere between $50,000 and $70,000 a year, worked so much unpaid overtime that, were he paid for it, he would stand to gain an additional $15,000 to $20,000 annually (Richtel, 2005b). EA employees also get no compensation time for their overtime, which prompted lawsuits (ea_spouse, 2004a, 2004b). The first was filed in California in July 2004, with a second later that year. A similar suit was also filed against Sony Computer Entertainment (Richtel, 2005b).

These lawsuits have caused workers to question the computer industry practice of rewarding employees with stock options and bonuses. Bonuses at EA range from 5 to 30 per cent of employee salary, though as studies show, these are not distributed uniformly among employees and are subject to crises in the larger economy (Shirinian, 2012). According to the managers, paying overtime would change the industry from one that values entrepreneurship to one where employees simply punch a clock. EA's employee benefits include an onsite gym, flexible work schedules and onsite amusement facilities, which include basketball courts, pool tables and, of course, video games. Also available for employee health were a masseuse and acupuncturist (Richtel, 2005b). One result of the complaints levelled by ea_spouse was the creation of an organisation geared to monitoring industry practices, called Gamewatch.org (ea_spouse, 2004b). Beyond this, however, these lawsuits raised the question of unionisation in an industry that had always resisted unions. Part of the difficulty centres around EA's use of a standard Silicon Valley measure of productivity: revenue per employee.

As a measurement, revenue per employee is problematic. As Table 5.7 shows, however, this measure skews the value of companies, allowing even small outfits to appear disproportionately valuable. A firm like Atari, when compared with a larger concern like EA, appears more productive to investors because it has a smaller number of employees, even though its revenue is also smaller. This method of appraisal allows an industry which produces a small part of a nation's GDP to appear much more instrumental, even as its jobs are being moved overseas.

EA claimed a $1 million per employee rating in 2004, but this would change drastically if overtime and comp time were factored in. Comp, short for

	Employees	Revenue (millions of US $)	Revenue per employee
EA	6,100	3,129.0	512,950.0
Atari	492	395.2	803,252.0
Microsoft	57,000	39,788.0	698,035.0
Sony	162,000	66,923.0	413,104.9
Disney	129,000	30,752.0	238,387.6
Time Warner	84,900	42,089.0	495,747.9
General Electric	307,000	151,300.0	492,833.9
General Motors	324,000	193,517.0	597,274.7
Monsanto	12,600	5,457.0	433,095.2
Halliburton	97,000	20,464.0	210,969.1

Table 5.7 Revenue per Employee Comparison, 2005
Source: Hoover's (2005).

compensatory, time is, essentially, time off in exchange for overtime worked. EA, however, has stated that, if workers demand too much, it would have no choice but to find new sources of labour outside California and possibly even in another country with cheaper labour costs (Richtel, 2005b). The company already has a major studio system in place in Vancouver and has already reduced productions at its West Los Angeles studio, which had opened in 2000 and produced a number of games tied to Hollywood licensing, including *The Lord of the Rings: The Battle for Middle Earth* (2004) and *GoldenEye: Rogue Agent* (2004) (Dyer-Witheford and Sharman, 2005; EA, 2005; Fritz, 2005b). In spite of the success of these Hollywood games, the company laid off sixty workers and was down to 320 developers and fifty individuals working in music, marketing and mobile content as part of this reduction at the Los Angeles studio (Fritz, 2005b). Despite calls for unionisation, none materialised, though in 2006, a number of EA workers succeeded in a lawsuit against the company that earned a collective US $14.9 million and promise of reforms across the industry (Huntemann, 2010). Erin Hoffman, revealed to be ea_spouse, went on to join the International Game Developers Association's (IGDA) board of directors and its Quality of Life interest group (Chalk, 2010b).

Because of the success of that lawsuit, there was some surprise when in 2010, a similar situation came to light with software developer Rockstar, best known for its *Grand Theft Auto* and *Max Payne* (2001) series of games (Huntemann, 2010). Rockstar, owned by game publisher Take-Two Interactive, served as a contrast to EA, as the developer drew on the successes of previous developers and on key franchises for its output rather than on the Hollywood model. Table 5.8 provides a snapshot of Take-Two Interactive's major holdings, including its properties owned via Rockstar Games.

In January 2010, the spouses of several developers at Rockstar published a letter on industry website Gamasutra.com complaining of poor working conditions including excessive crunch time and declining benefits (Chalk, 2010b). Complaints focused on a long-term state of crunch time as the company struggled to complete *Red Dead Redemption* (2010), the sequel to its successful *Red Dead Revolver* (2004) game. Reports suggested that crunch time began in March 2009, with employees working twelve-hour days, six days a week (Chalk, 2010b). In a few days, other complaints arose – typically anonymously and online – suggesting that the company not only threatened loss of bonuses but layoffs should deadlines not be met as well as loss of credit for work (Bertz, 2010). The question of who gets credited for work has a long history within the industry. Early games programmers were often denied any onscreen credit for games, while in later cases, leaving before a game was put on the market might result in withdrawal of onscreen credit. In an industry that relies on reputation

Year formed	1993
Headquarters	New York, US
2012 Sales (US$m)	825.8
2012 Employees	2,235
Industry sector	Software publisher
Key software franchises	*Grand Theft Auto*
	Red Dead Revolver/Red Dead Redemption
	Max Payne
	Midnight Club
	Dora the Explorer
	Go, Diego, Go
	BioShock
	Sid Meier's Civilization
	Bully
	Manhunt
Studios owned	Rockstar – New York, US
	2K Games – New York, US
	2K Sports – New York, US
	Venom Games – Newcastle upon Tyne, UK
	Firaxis Games – Hunt Valley, Maryland, US
	2K Boston – Boston, Massachusetts, US
	2K Australia – Canberra, Australia
	2K Marin – Novato, California, US
	2K Czech – Brno, Czech Republic
	2K Shanghai – Shanghai, China
	Gotham Games – New York, US
	Jack of All Games – West Chester, Ohio, US

Profile

New York-based Take-Two Interactive publishes software for consoles and personal computers. Its brands have been so popular that it was recently the target of a takeover bid by rival Electronic Arts (Take-Two, 2008c). It counts among its important franchises both the *Grand Theft Auto* series, acquired with its purchase of Rockstar Games, and a number of sports titles, including the ESPN franchise, before it was taken over by Electronic Arts (Hoover's, 2008). The company also has branches in Switzerland and Canada, Japan, the UK, Australia and Austria (Take-Two, 2008a) In 2007, following a number of legal scandals, an almost entirely new board of directors was installed, giving the company impressive ties to a number of other major operators. It has ties to Blockbuster, Inc., Columbia Pictures, the MessageClick, Inc., News Corp, MCI, BMG, the William Morris Agency, Major League Baseball and the World Wrestling Federation (Take-Two, 2008b).

Table 5.8 Corporate Profile of Take-Two Interactive Software, Inc., 2012
Sources: Hoover's (2012); Mergent (2008).

and a history of projects to help secure future work, loss of credit could have serious consequences. Because Rockstar was a sizeable development studio, with branches around the US, employing roughly 900 workers, the complaints made headlines (Totilo, 2010). The company's response was to deny many of the claims and suggest employees talk with the human resources department (Chalk, 2010a; Totilo, 2010). Eventually the complaints seemed to die down, without lawsuit or any explanation of how the situation might have settled. *Red Dead Redemption* shipped late, on 18 May 2010, and shortly after the company laid off approximately forty workers from the San Diego division, which had developed the game (Graft, 2010b).

Analysts have pointed to these struggles, particularly over crunch time, as one possible reason why employment in the software side holds little appeal for women. Such schedules rely on a relatively young labour force, with limited responsibilities outside the job (Huntemann, 2010). Even with these continued labour problems, few experts predicted any success in unionising the industry from within. One union organiser, from WashTech, has been trying to unionise high-tech workers for years, but has only attracted 450 members between 1998 and 2005 (Richtel, 2005b).

CASE STUDY: HOLLYWOOD UNIONS, VIDEO GAMES AND LABOUR

A second labour problem has presented itself. As discussed in Chapter 4, the video game industry has formed close ties with Hollywood, relying more and more on film content. At the same time games have also become more dependent upon Hollywood talent, particularly for voice work. Again the video game industry must ask itself how to deal with unions.

In 2004, almost 2,000 unionised actors found work in video games, including such notables as Ewan McGregor, Toby Maguire and Willem Dafoe (Gentile, 2005a). As game profits have risen, actors have begun to demand a bigger share of the pie. Recent contract negotiations between the industry and the Screen Actors Guild (SAG) and the American Federation of Television and Radio Actors (AFTRA) focused heavily on ensuring that voice actors for video games received compensation including health care, pensions and residuals (Brodesser and McNary, 2005). Unlike film and TV, however, the video game business has no history of negotiating with unions or of paying residuals to anyone involved in game creation (Brodesser, 2005). Because games' profitability is less than that of films, residuals were a hotly contested issue in the negotiations. The previous contract provided minimum pay rates with no provision for residuals, in part because so much of video game voice work has always been done in-house (Brodesser, 2005). The industry is also wary of residuals because currently only 10 to 15 per cent of all games involve unionised workers, and those workers are typically tied to other industries such as film (Fritz, 2005f).

The negotiations were also difficult because representatives of the video game industry had come together informally, rather than as a unified group (Brodesser, 2005). Ultimately, a contract was ratified between those representatives and AFTRA, calling for a 36 per cent pay hike but granting no residuals (Fritz, 2005f). SAG's negotiating committee endorsed the same deal, but the general membership voted it down. This will probably limit future negotiations with SAG, making AFTRA the powerhouse union to deal with video game work (Fritz, 2005g). In fact, a contract was later ratified, negotiated with AFTRA, detailing the same 36 per cent pay hike, greater contributions towards retirement plans, shorter periods of work required to earn overtime pay and other benefits (Bank, 2005). Part of SAG's worry was the unchecked ability of the video game industry to act like movie studios and television networks, rather than simply as publishers (Brodesser, 2005). As discussed in Chapter 4, this tendency is only likely to increase and will no doubt result in more heated negotiations between unions and the industry in the future.

Part of the trouble faced by union organisers is that union membership has been declining across media industries, and had never taken off in video games to begin with. As noted earlier, on the hardware side, most labour is not unionized, in part because companies have sought countries with less stringent labour rules, while the software side is modelled after the notoriously anti-union computer software industry. In some industries, like telecommunications, union membership has declined by as much as 50 per cent since 1985 (Grover *et al.*, 2005). As these industries continue to expand internationally, the difficulties of organising seem likely to increase.

THE PLAYER CHALLENGE TO LABOUR

Finally, the video game industry is also sensitive to questions of player-centred labour. Player-centred labour refers to the value added to games through player production. This might include the value simply of having a large player base, but more commonly refers to player production of virtual goods within a game. It can also refer to content that builds onto the game or content that adds information outside, such as game Wikis or community bulletin boards, etc.

One genre which particularly benefits from player-centred labour is that of massive multi-player online role-playing games, where players contribute to the play environment. In 2004, there were almost 350 such games, with more than 10 million players (Wallace, 2005). In these games, players often create virtual goods for trade, and these goods have made their way into real markets via online services like eBay (Castronova, 2002, 2005). For instance, a player of the hit game *Ultima Online* earned more than $25,000 by trading in goods he created in the game. Estimates suggest that the real-world value of these virtual goods and services is almost $880 million per year, not including the cost of the games or subscriptions (Wallace, 2005). The industry has tried to enforce the idea that any such property belongs to the company, but this has increasingly been challenged by governments and gamers alike (Klang, 2004). In Beijing, the government has begun to limit the use of virtual currencies, while South Korea has started to tax virtual goods and currencies (Barboza, 2009; Rosenberg, 2010). The massive multi-player platform – whether it is a game is contested – *Second Life* (2003), has taken a different approach, allowing players full intellectual property control of what they create in the game. Estimates suggest that an average player's transactions are worth almost $1,000 a month, and this amount is increasing roughly by 25 per cent per month. One player even claims to have made $100,000 per year in the game's real-estate business (Wallace, 2005).

The value of virtual goods has grown so much that a market has emerged to satisfy it and a new labour practice, known as 'gold farming' has been created to address the need. Virtual trading of goods has become important in MMORPGs like Blizzard Entertainment's *World of Warcraft* and CCP Games' *EVE Online* (2003). Gold farming relies on workers playing games for hours on end, accumulating gold that can be sold outside the game, using services like PayPal, for real money (Nolan, 2010). Players who purchase the gold can then use it to buy virtual goods they can't acquire without massive increases in their own playing time. According to a 2011 World Bank study, gold farming has become an industry in itself, worth roughly US $3 billion (Shapira, 2011). In games which feature their own in-game currencies, such as *EVE Online*, *World of Warcraft* and even *Second Life*, trade for those currencies has begun to have a real-world impact. Researcher Edward Castronova discovered as far back as 2001 that virtual economies could be compared to real-world ones. Playing the Sony MMORPG *EverQuest*, he estimated the game's Gross National Product (GNP) per capita placed it between Russia and Bulgaria, while its currency was valued higher than many real-world ones including the Yen and the Lira (Castronova, 2001; Coutts, 2009). Some countries have begun to worry about these virtual economies; in 2009, China introduced regulations to prevent trading in virtual currencies (Barboza, 2009). Their fears aren't unfounded, as players have begun to take advantage of virtual economies and gold farming has frequently been compared to sweatshop labour (Coutts, 2009). Gold farming has become big business, particularly in Asia. Gold farming businesses are burgeoning in China, particularly in Beijing and Changsha. The World Bank study indicates that as much as 70 per cent of revenues from Asian gold farming stays within the region, though workers earn an average wage of US $2.70 an hour (Shapira, 2011).

Virtual worlds live and die based on the value added by players, so player labour becomes an important contributor to their persistence. This amounts to free labour or, in the case of gold farming, low-cost, externalised labour costs that are vital for the industry, particularly to the survival of smaller developers, who might also have to rely on free testing to bring their games to completion. Because of how much time players have spent working, the reaction to changes in these unintentional markets has prompted demands for a change in industry practice to allow players some say about, and potentially profit from, content they've created (IGDA, 2004a). Industry response has been mixed, with some companies trying techniques similar to those of *Second Life*, though rarely granting full control over property. These games are particularly popular globally, and so it may be that the US has to take its cue from foreign countries and game designers, such as Chinese company Shanda Entertainment, which has created its own eBay-like market for players to sell characters and game-produced goods, with Shanda taking a cut of the profit (Grover *et al.*, 2005).

EDUCATION AND THE VIDEO GAME INDUSTRY

One factor not included in these labour surveys is that of education. This is surprising because education is typically one of the defining features of employees in information industries, and also because education programmes targeted specifically at the industry are being created. As previous chapters have indicated, the game business is beginning to attract government support in many countries. One way this has been instituted has been through education. In the US, a number of schools, including the University of Southern California, Massachusetts Institute of Technology, the University of Washington and Carnegie Mellon University, have initiated video game design courses and programmes (Loftus, 2003). As the technologies involved have become more affordable, it has become possible for even small universities to incorporate these subjects into their programmes (Wallace, 2004). While the earliest programmes emerged in the US and have received the most attention, it must be noted that similar programmes have emerged worldwide.

These programmes tend to be professional in nature and heavily interdisciplinary. For example, Carnegie Mellon University's Entertainment Technology Center draws faculty expertise from both the University's fine arts and computer science departments (Deutsch, 2002). While a majority target undergraduates, graduate options are also available. The first Master's programme was at the Rochester Institute of Technology in New York, but both Georgia Tech and USC now offer Master's programmes in game design (Deutsch, 2002; Schiffmann, 2002).

Video game design programmes tend to be small. Southern Methodist University accepts only 100 people per year for a course which will take eighteen months and cost students US $37,000 in tuition (Carlson, 2003a). Some programmes are even smaller. The DigiPen Institute of Technology in Washington, sponsored in part by the industry, graduated eleven students in its first year, and only thirty-six in 2001 (Schiffmann, 2002). The programmes also reflect the industry's gender breakdown, attracting predominantly male students. This is the case at the Institute of California, in Orange County, where only twenty-five out of 150 students are female (Swett, 2003).

The industry, too, has taken the idea of games education seriously. The larger software publishers have leveraged their power to help support particular programmes. These ties have also allowed the major companies to access college students for research and marketing purposes. Electronic Arts, for example, sponsors college advertisers and what they term 'guerrilla marketing tactics' to help promote their games to students. (Miller, 2005). Each year, *Game Developer*

Country	Number of programmes		Country	Number of programmes	
	2005	2012		2005	2012
Afghanistan	N/A	1	Italy	1	2
Argentina	1	2	Japan	2	2
Australia	9	25	Mexico	2	7
Austria	2	4	Netherlands	2	4
Belgium	1	1	New Zealand	3	2
Brazil	4	4	Norway	1	2
Canada	38	69	Pakistan	1	1
Chile	1	1	Philippines	N/A	1
China	1	3	Portugal	N/A	2
Colombia	N/A	1	Serbia	N/A	1
Denmark	3	4	Singapore	4	4
Finland	1	3	South Africa	N/A	2
France	3	4	South Korea	2	1
Germany	3	12	Spain	3	5
Great Britain	38	51	Sweden	11	13
Greece	1	1	Switzerland	1	2
Hong Kong	3	1	Taiwan	N/A	1
India	2	13	Thailand	1	1
Indonesia	N/A	1	Turkey	1	2
Ireland	2	3	United States	169	296
Israel	1	1			

Table 5.9 Number of Programmes with a Video Game Studies Component by Country, 2005 and 2012

Sources: Game Career Guide (2012); IGDA (2005).

Magazine publishes a career guide for the industry and sponsors surveys on employment. One concern expressed by the industry is how well such education actually prepares students for jobs (Moledina, 2004b). Such programmes may face problems as the industry is experiencing not only rising labour dissatisfaction but also high levels of outsourcing Table 5.9 provides a comparison of the number of schools offering video game studies by country in 2005 and 2012. While the data are largely self-reported, it still furnishes a useful snapshot of the increasing attention being paid to video games in academia. It clearly shows both a tremendous growth – with approximately 320 programmes in 2005 to more than 530 in 2012 – and also that the largest growth is in the regions where most video games commodities are consumed: North America, Western Europe and Japan.

CONCLUSION

The nature of employment and production within video games is a confusing area. On the one hand, the labour involved is an ideal example of information labour, as it is creative, requires a high degree of education and skill and relies heavily on the production and manipulation of information itself. It must be noted that virtually all studies of labour in the computer and video game industries have focused on software production rather than hardware. On the other hand, however, it exhibits a number of troubling patterns, including unequal gendered labour and employee dissatisfaction, particularly in regards to workload and crunch time schedules.

In part this owes to the time and money it takes to produce a video game. This pattern of creation, tied to periodic buying seasons, means a built-in crunch time for workers, even as it suggests possible limits with products rushed to meet deadlines. Some important steps have been taken to help deal with this, but they have not always been welcome ones. Licensing concepts from other media, for example, may allow for decreased production time but limits worker creativity, one of the elements drawing people to jobs in the industry.

An important thing to realise about the production of video game commodities is that they are heavily driven by the value added by highly skilled employees. The overall cost of manufacture for video games comes almost entirely from labour expenses. This makes the labour market a vital factor in understanding the industry. Workers meet the criteria of information labourers but lack the power in either markets or production suggested by theories of the information society. On the software side, labour is highly skilled and creative and, as such, is seen as in demand and worth promoting by government. On the hardware side, manufacturing jobs that are seen as less creative and more likely to unionise tend to be exported to regions where labour and environmental conditions are more favourable.

The study of the video game industry confounds the myth that information labour gives workers more power and satisfaction. As studies have shown and the case studies in this chapter attest, workers are subject to heavy workloads that often result in their leaving the industry altogether. Not only do studies show that an increasing number of employees experience decreasing job satisfaction, but they also show that, as the economy falters, the perceived benefits of such employment are progressively eroded. The fall in the value of stock options and bonuses following the dot-com collapse of the late 1990s is only one example of this.

The gendered patterns of employment are also significant, with the most technical and creative positions not only going to males, but often with significant wage disparities. Combined with the skewed gender representation, particularly on the software side, it could be argued that the desire to market more effectively to wider audiences may suffer due to lack of representation within the industry.

Both of these concerns are related to the current labour disputes, which have brought the question of unionisation to a sector of the economy known for its anti-union sentiments. Although attempts at unionisation have thus far been ineffective, this may change in the future because video games are strengthening ties with already unionised entertainment industries such as film, television and recorded music. The potential for unionisation faces one other limit in the increasingly globalised production, with hardware manufacture jobs moving away from the sites of games consumption and software jobs under threat by the same practice. As the process of game production becomes more obscure, public support for unions is even harder to get, with firms able to continue to relocate whenever labour attempts to organise. This strategy has proved particularly difficult for skills-based unionisation to overcome, perhaps calling for a change in union tactics and organisation to more adequately respond to high-tech industries.

Finally, the question of what it means to work in video games is influenced by the challenge of value added labour by players. While player added value is largely limited to MMORPGs, as the sophistication of game consoles evolves and the convergence of technologies into consoles and portable devices allows them to become media hubs, the problem may grow thanks to the increased range of ways in which players can create and manipulate content. As players begin to interact with games in various ways and exploit the social capabilities accompanying networked machines, the more chances they will have to impact game content. This presents the question of ownership and compensation already being raised over virtual goods. Addressing these questions is one of the big challenges for the industry in coming years.

Conclusion: Making Sense of the Global Video Game Industry

Over the last forty years, video games have gone from being a small-time fad to representing a successful cultural industry with a unique and diverse audience. Like any cultural industry, the global video game business has changed since its earliest days. These changes have revolved around the adoption of particular rules – or logics of production – that govern the way in which its commodities are created. It is hoped that understanding these logics can shed some light on the more common controversies surrounding games. To question what should be done about violent games without asking why they were made could mean that we overlook possible solutions. One of the goals of this study has been to describe this industry and its logics of production.

But no business functions in a vacuum; in the last twenty years, numerous links have been forged with other cultural and media industries. Because certain limits impinge on the success of video games, the business has taken care to foster such links. With game popularity galvanising a rise in game studies, and academic programmes, an understanding of how the industry and its processes relate to other forms of communication is important.

The business model draws on the computing, toy and, especially, film industry, adopting particular practices from each. Chapter 1 demonstrated that this evolution was not accidental. Rather, video games actively tried to mirror tried and tested systems' logics of production. The major sectors are hardware manufacture, software development, software publishing and retail.

In the Introduction, the logics that govern industry decisions were discussed. These logics were:

- ensure and maximise profit
- keep production costs and, therefore, risk down
- ensure long-term demand by creating long-term audiences
- when possible, control demand
- make use of planned obsolescence to drive demand and prompt innovation
- make products that garner attention
- maintain industry autonomy and stability.

As the economic history of the industry has been laid out in addition to the structure and ties of the modern industry, it is clear that the business has been developed with these logics in mind. One key way these goals have played out has been through a concentration of power.

Power has been concentrated in hardware manufacture and software publishing. Four companies, Sony, Microsoft, Nintendo and Electronic Arts, dominate, accounting for a majority of revenues. The first three represent both hardware manufacture and software publishing, while Electronic Arts has managed to achieve its success based solely on software publishing. Currently, most of the revenue comes from software, thanks to an intense competition between the three console makers that has driven them to drop the prices to a point where they gain little profit from them. The concentration of power maintained by Sony, Nintendo and Microsoft has allowed the companies to insulate themselves from market shocks, regulatory demands and efforts from other sectors to influence how the industry functions. Because all three manufacture both hardware and software, they are less subject to demands from the software sector over control of content and labour practice.

When negotiations between these sectors does occur, it is typically through strict licensing deals, with hardware manufacturers dealing with software publishers rather than smaller developers. Again, this has served to restrict the potential of small companies and their workers to exercise power within the industry. This has led to increased consolidation in which software publishers have bought smaller developers. While risk has been reduced for these companies, for others, it has grown. The consolidation of power within the industry has had a deleterious impact on small software developers, who are forced to negotiate with publishers and hardware manufacturers for each game they create. It has also resulted in greater reliance on licensed content and franchised games, which means innovative games have a harder time reaching the market. In contrast, the retail sector has had to deal with both hardware and software manufacturers. It remains largely unaffiliated through ownership or licensing with either hardware or software sectors and, while still concentrated, there remains room for small vendors to enter the market. As digital technologies have advanced – particularly with regard to mobile phones, wireless capabilities and increasing Internet bandwidth – challengers have begun to present themselves, threatening to alter the existing dynamic.

The typical production time for a video game is increasing, now taking as long as two years. As the development time for games has grown, so has the cost. Because of this, the industry seeks to exploit licensed content from other media, thus minimising risk and taking advantage of existent marketing and advertising.

The production of games is driven by the capability of the hardware. The hardware sector has worked on a system of two- to three-year periods of planned obsolescence. As hardware evolves, the demands and costs of software development rise as well. Development involves four main phases, with the labour required increasing as the project nears completion. It is at the end of the process that localisation of products occurs. It is also during this end period that the 'crunch time', at the root of much worker dissatisfaction, occurs, with employees often called upon to work eighty-hour weeks with no days off and no compensation for overtime.

The promotion of games has become a major part of the process, with budgets often equalling those of production. As with toys, video game production and marketing are focused on the Christmas buying season, and to a lesser extent on tie-ins with Hollywood's summer blockbuster films, though this relationship has begun to shift as games have moved from an ancillary market for films to a rival for screentime. Thus the majority of advertising dollars is expended during the holiday season.

Distribution of products is handled by the major companies – a negotiation between the hardware manufacturers and the software publishers. As with DVDs and recorded music, large general goods stores like Wal-Mart and Best Buy make up a significant part of the retail process. Rental of games is on the rise, and here concentration is similar to video and DVD rentals. At this time, the major companies have expressed little interest in gaining control over the retail side, but this may change in the future, particularly as the cost of development rises and as digital distribution becomes increasingly viable.

The industry is also highly globalised, relying not only on international audiences, but on a system of production and ownership which is also international in scope. Not only are the major markets for video game products centred in North America, Europe and Japan, but these are also the primary locations for software production. The industry's rapid growth and high profitability, combined with its reliance on a form of skilled labour viewed as highly desirable, attracted considerable government and educational attention, with many countries and regions eager to lure production their way. Examples include the UK, Australia, a number of regions in the US and many countries in Eastern Europe. These initiatives have often been tied to attempts to draw film production as well.

Industry studies have shown that, while globalisation has kept production costs down – often by locating where labour and environmental regulations are lax – job satisfaction isn't as high as might be expected. Those same studies also indicate that most employees leave the industry fairly quickly, often after fewer than seven years. Moreover, labour is bifurcated, with hardware production occurring in countries that are not the primary markets for video game products. The work involved in hardware creation is often hazardous and environmentally damaging, but this is a largely hidden cost to the creation of video games.

The commodification process is a means of making social relationships concrete. Producing video games as commodities creates items for exchange, but eventually also means that the way in which they are created and how they are used come to be seen as a natural state. At the same time, the relationship between labour and management undergoes a similar process. The production of video games as a commodity has evolved into a system in which dissatisfied labourers come to view their dissatisfaction as natural and unchangeable. A second relationship is also cemented by the creation of the video game commodity: the relationship between the consumer and producer. In this case, expectations become fixed about which types of games should be made and when. Costs, planned obsolescence and types of games seem natural. This results in the view from outside the industry that video games function neither as art, nor should they be considered acts of political expression. They are simply playthings with nothing to say, unworthy of consideration.

At the same time governments around the world are struggling to deal with issues of public perception about problems (such as violent content). Regulation is primarily accomplished through self-regulation. In the US the rating system for games is one of the most stringent of any communications industry. Retail associations have responded to consumer pressure by increasing regulations and punishments for retailers who sell games to inappropriate individuals. There is still intense scrutiny from various state sources, both within the US and outside. Ratings systems outside the US tend to be even stricter, with particularly complex examples prevailing in the European Union, Japan and New Zealand. Though these systems tend to be more stringent, the industry believes them confusing and likely to result in unfair restrictions on content that the industry would prefer to see decided upon by the market. While investigations initially focused on violent content, increasing attention is being given to workers' rights and to matters of intellectual property.

While the primary commodities are video games, these are no longer the only way to ensure and maximise profitability. There is increasing focus on licensed products related to games, including specialised hardware. These products, however, are typically licensed to smaller producers, who shoulder most of the risk.

As the industry has matured, other product lines have emerged. Increasingly, games are licensed to toy manufacturers and other media avenues for conversion into film and television. Such products are prone to becoming advertising-driven.

In television and film, this has resulted in a dependence on formats seen as most reliable, such as the action film, at the cost of innovative content. It is not hard to imagine such a fate befalling video games. Typically, these licensing deals have originated from outside industries. But this balance has begun to shift and, as it does so, new struggles are likely, with new forms of labour such as voice acting, which have unfamiliar histories of organisation and governance. Internal franchises are becoming increasingly important, and the desire to acquire popular franchises has been one of the key reasons for consolidation within the industry.

More recently, the industry has developed its second commodity: the audience. This is similar to the dual-product market formula characteristic of many media forms, particularly television, and forces the industry, albeit not unwillingly, to consider the role of advertising in games. For some genres, such as sports, this shift will be less problematic, but for others, it may prove intrusive. The impact of advertising has been modest in comparison to other industries, but it is a growing source of revenue, particularly as mobile gaming takes off and new audiences are courted around the world.

Studies have shown that more and more people in the eighteen to thirty-four demographic, one of the most important groups for advertisers, are shifting their time away from television and other media to video games. The cultivation of this demographic has prompted increased interest from advertisers. Currently, the majority of game revenue comes from sales, but it is expected that advertising revenue will become a major source of income in the future. Experiments incorporating advertising into video games are already being conducted. The most problematic form of these is the advertiser-driven game. This rise of advertising is likely to result in further insulation of the major companies from consumer demand as well as renewed questioning of advertising's impact on content.

Even from its earliest days, video games relied on ties with other media sectors, its links with the movie business extending back into the late 1970s, when film companies first began to explore the possibilities of video games as licensed properties. The relationship has not always been beneficial, however. Both industries experienced losses in the process. Setbacks hit games first, almost ending one of the major companies, Atari. Atari's volatile history illustrates the struggle between profitability and control the industry has faced in the past when working with other sectors. For example, Hollywood's bid to exploit the popularity of video games has been a rocky road, involving buying into and then selling out its interests in video games in both the early 1980s and the mid-1990s. Often attributed to a clash of cultures, this also reflects tussles over control of ideas and production processes.

As the game industry stabilised, it again became attractive to other media concerns. When games revenue recently surpassed Hollywood box office revenues, it was seen as a signal that the business was around to stay and the relationship between video and film vouchsafed mutual advantages. Video games have become important to other industries as well. Both television and recorded music have experimented with licensing their products through video games. The advantage of these ties is threefold. First, the creator of a brand could earn additional profits. Second, the product of one industry could piggy-back on the marketing of the other. Third, it has been seen as a means to minimise the risk of releasing products for all parties. The downside of this alliance has been an overreliance on licensed content in spite of consumer preference, as discussed in Chapter 4. Industry concentration and strategic alliances with other industries has served to minimise the industry's responsiveness to consumer and labour demand in the market.

The impact of video games on other media cultures goes beyond their licensing ability. In part this is thanks to the fact that technologies are brought together in products that operate on a variety of platforms and can mimic the functions of various other products. The release of the eighth generation of consoles by Sony, Microsoft and Nintendo is expected to bridge the gap between video games, television and personal computers among other devices. In part, Microsoft's foray into games can be seen as an attempt to ensure its influence over as many computer devices as possible. Two other examples of this convergence are evidenced in the growing ties between video games and the mobile phone and online markets, enabling video games to draw audiences away from other media. However, it must be stressed that the industry has maintained a considerably broader demographic, attracting increasing numbers of older players and female players. As this has happened, other types of content have been explored. Casual gaming, in particular, has helped to tap into these demographics while also pushing convergence with cell phones and other portable media. This convergence of technologies has drawn interest from major companies. Microsoft's entrance into the market has been an attempt to create a media hub under its complete control. The potential for video game units to become the centre of household media has sounded alarms for other media industries.

The other major concern has been the increase in player-created content. This value added problem is particularly crucial for online role-playing games in which players devote increasing hours to creating virtual goods. These goods, which often include players' avatars, have started to seep into the real world. Items are being traded on eBay, and some companies are beginning to work on ways to negotiate the ownership of these goods. The intellectual property implications of this are still being explored, and producers of the multi-player games where these creations most commonly occur are likely to continue actively pursuing control of player-generated content. Video games also pose other problems to other media industries. The increasing sophistication of their technology, coupled with the frequent inclusion of game engines, has resulted in unexpected consequences for other industries. Players can now produce their own movies using tools included in their games. This development has been dubbed 'machinima' and is drawing attention from film-makers and advertisers. While both video games and films have begun to try and incorporate this content into profit strategies, it still remains an area of considerable concern.

Increasingly, the video game industry is developing international markets for its products. However, most markets are currently in industrialised countries, with the business working to maintain distinctions between these regional markets. This has been accomplished in two ways. First, products developed for one platform in a particular regional market do not work on platforms in another, a tactic also seen in DVDs. Second, there is a growing tendency to localise a product, tailoring a game's contents to local attitudes and practices. Product markets have focused on platform developments, and as the industry seeks out ties with other media forms, it is incorporating a variety of new capabilities such as the ability to play music, display photographs and access the Internet.

As Chapter 5 showed, one of the most crucial commodities for video games is labour itself. While the business seems the ideal example of an information industry, the struggles workers face both on the hardware production side and in the more commonly studied software side suggest that information labour brings with it a number of problems. In an information industry, labour is seen as having significantly more power than in other spheres. Labour in the video game industry exhibits a number of interesting characteristics. Theories of the information society would suggest that labour should be highly satisfied, educated, mobile and well compensated. Studies suggest that not only is employee dissatisfaction on the rise, but that compensation has been limited to models based on the overarching computing field. Worker mobility is increasingly limited by consolidation and transnational production practices. Because the development of software is structured with most major producers owning or licensing with small, production groups, employees need to work harder to ensure they will continue to find new projects. It is significant that workers are expressing increasing job dissatisfaction and often view the industry as a stopover on the way to more stable, lucrative careers. The rise of game studies programmes in universities is seen as one way to address the continuing need for labour. The high levels of industry concentration will continue to limit the effectiveness of labour to influence the production process.

Beyond this, however, there is a distressing disparity in terms of gender. In this sense, the video game world is similar to computing. Fewer than 10 per cent of employees are female and, in almost every career track, females are paid less money despite similar levels of experience. The industry is also dominated by younger workers, on the lower end of the pay scale.

Contrary to what theories of the information society would suggest, the best-paying jobs are not the creative, highly skilled positions. Studies indicate that employees working in business and legal consistently make more, even as they account for a smaller proportion of workers. More telling is the fact that employees engaged in creative work – the programmers, those in audio and other fields – are increasingly treating video games as a stepping stone. As this happens, fewer and fewer employees from those areas are reaching the higher pay levels contingent upon experience.

This desire to find more stable jobs elsewhere is partly due to the industry's measure of its own success by revenue per employee. Such a measure overvalues the industry's production, as compared to other companies and businesses, allotting disproportionate value to smaller firms with even a single big hit without accounting for other debts or fluctuations in audience demand. Measured this way, the success of small companies is overvalued while the failure of larger companies is deemed the fault of workers because debts, market fluctuations, management decisions and other factors out of individual workers' control aren't taken into account. As such, management must seek to drive employees to work longer hours at lower costs in order to guarantee a favourable productivity measurement.

Because labour makes up the bulk of production costs, as game complexity increases, so will costs, a reason behind the reluctance to change wage and benefit packages. In spite of this, unionisation is not seen as an option by most people, including workers and managers. This may change, however, as video games become increasingly tied to licensed Hollywood products and Hollywood labour, which is highly unionised. Of course, unions themselves are embedded within historical contexts. Present union formulations may not work well with the video game industry, but a shift in organisation – from trade unions to industrial unions, for example – might allow unions more success in gaining members from both video games and other high-tech sectors.

These challenges set an interesting trajectory for the future. While continued growth in profits has been predicted, where those profits come from is increasingly up in the air. Having focused on the North American, Western European and Japanese markets, the industry has created a stable audience. How to best address the remainder of the world, where there is clear interest video games, is a question the industry is only beginning to answer. In addition, the advent of new technologies and ways to play have already begun to destabilise the industry's long-standing power dynamic. Digital distribution has the potential to reshape affairs, and challengers to the existing structure have emerged both within and outside the business. The industry has already begun to try to cater to a variety of players, rather than just the hardcore gamer so often featured in media and in academic studies.

In this regard, understanding industry practice poses a challenge to anyone reporting on, studying or criticising video games and the industry which creates them. There has been a tendency to focus on particular products and easy narratives – video game violence and addiction, the rapid growth of the sector and how it might offer productive new areas of employment are just some examples – that misses the complexity of games, their production, impacts and potential. Much of this focus has treated the products, production and impacts of video games as both inevitable and fixed. This ignores the fact that the video game business and its products are the result of choices: choices about the audience, choices by the audience, choices about how to structure the industry and reinforce relationships within it, choices about how it should interact with other industries and choices about what questions to ask about video games. Understanding and better utilising such complexity is necessary for not just understanding video games but for better policing them and fostering the industry that produces them. Just as the industry's structure has changed, so, too, can its products, their impact on our lives and the ways they are policed. Failure to acknowledge the element of choice on the part of industry, audiences and critics results in a failure to understand the complexity and potential of games themselves. To miss those moments risks missing the true potential of what video games have been and what they can be.

Bibliography

Abraham, Linda Boland, Marie Pauline Mörn and Andrea Vollman. (2010). 'Women on the Web: How Women Are Shaping the Internet' (White Paper), ComScore.

Activision. (2004). *Annual Report, 2004*. Santa Monica, CA: Activision, Inc.

Activision. (2007). *Annual Report, 2007*. Santa Monica, CA: Activision, Inc.

Activision. (2008). 'Activision-Blizzard: Overview', http://investor.activision.com/ (downloaded 27 July 2008).

Albrecht, Chris. (2008). 'HDTVs in Nearly a Quarter of U.S. Homes', 11 December, http://newteevee.com/2008/12/11/hdtvs-in-nearly-a-quarter-of-us-homes/ (downloaded 15 September 2010).

Alpert, Bill. (2003). 'ATI Zapping Rival Nvidia in Graphics Chip Wars: Beating Expectations'. *National Post*, p. IN.1.Fr., http://proquest.umi.com/pqdweb?did= 358704351&Fmt=7&clientId=5258&RQT=309&VName=PQD (accessed 15 July 2013).

Anders, Kelly. (1999). 'Marketing and Policy Considerations for Violent Video Games'. *Journal of Public Policy and Marketing* vol. 18 no. 2, pp. 70–3.

Aoyama, Yuko and Hiro Izushi. (2004). 'Creative Resources of the Japanese Video Game Industry'. *Cultural Industries and the Production of Culture* vol. 33, pp. 110–29.

AP. (2002). 'Army Recruiting through Video Games', 23 May, http://www.nytimes.com/ aponline/technology/AP-Video-Games-Army.html (downloaded 23 May 2002).

AP. (2003). 'Schwarzenegger Promotes Atari Video Games', 14 May http://www.highbeam.com/doc/1P1-73914808.html (downloaded 15 May 2003).

AP. (2005a). 'Electronic Arts Signs 15-Year ESPN Deal', http://apnews.excite.com/ article/20050118/D87M2VG0.html (downloaded 18 January 2005).

AP. (2005b). 'Legislator Seeks to Restrict Teens' Access to Video Games', 22 February, http://usatoday30.usatoday.com/tech/news/2005-02-22-ala-game-ban_x.htm?csp=34 (downloaded 22 February 2005).

AP. (2005c,). 'Missouri Bans Video Games from Prisons', 24 January, http://usatoday30. usatoday.com/tech/news/2005-02-22-ala-game-ban_x.htm?csp=34 (downloaded 26 January 2005).

AP. (2005d,). 'Yahoo Launching Mobile Games Studio', 4 March, http://usatoday30. usatoday.com/money/industries/technology/2005-03-04-yahoo-cell-games_x.htm?csp=34 (downloaded 4 March 2005).

Aslinger, Ben. (2008). *Aural Appearances: Popular Music, Televisuality, and Technology*. Unpublished dissertation 3327830. Madison: University of Wisconsin Press.

Aslinger, Ben. (2009). *Genre in Genre: The Role of Music in Video Games*. Paper presented at Breaking New Ground: Innovation in Games, Play, Practice and Theory Conference, 1–4 September, London.

Aslinger, Ben. (2010). 'Video Games for the "Next Billion": The Launch of the Zeebo Console'. *Velvet Light Trap* vol. 66 (Fall), pp. 15–25.

Aslinger, Ben. (2013). 'Redefining the Console for the Global, Networked Era'. In Nina B. Huntemann and Ben Aslinger (eds). *Gaming Globally: Production, Play and Place*. New York: Palgrave Macmillan, pp. 59–73.

Atari. (2008). 'Atari – Corporate Profile', http://phx.corporate-ir.net/phoenix. zhtml?c=66845&p=irol-homeProfile&t=&id=& (downloaded 27 July 2008).

Australian Classification Board. (2008). *Classification Requirements for Computer Game Retailers*. Sydney: Australian Government Attorney-General's Department, Classification Operations Branch, http://www.classification.gov.au/ resource.html?resource=865&filename=865.pdf (downloaded 11 May 2010).

Australian Classification Board. (2010). 'Classification Markings on Film and Computer Games', http://www.classification.gov.au/Guidelines/Pages/ Guidelines.aspx (downloaded 11 May 2010).

Baar, Aaron. (2011). 'Survey Finds Phones Revolutionize Games', 1 March, http://www.mediapost.com/publications/article/145761/survey-finds-phones-revolutionize-video-games.html (downloaded 3 March 2011).

Bajak, Frank. (2005). 'Next-Generation Xbox to Be Media Hub'. *Associated Press*, 3 May, http://www.eastvalleytribune.com/get_out/article_7f99688f-37a9-5533-8950-515dc6bddb9f.html (accessed 13 May 2013).

Bank, Alan. (2005). 'Screen Actors Guild Approves New Game Voiceover Contract', 29 July, http://www.gamasutra.com/php-bin/news_index.php?story=6070 (downloaded 22 July 2005).

Barboza, David. (2009). 'Beijing Limits Use of Virtual Currencies in Gaming Industry'. *International Herald Tribune*, p. 13, http://ezp.bentley.edu/login?url=http://search. proquest.com/docview/318997955?accountid=8576 (accessed 15 July 2013).

Barboza, David. (2011). 'Workers Poisoned at Chinese Factory Wait for Apple to Fulfill a Pledge: [Business/Financial Desk]'. *New York Times*, p. B.1, http://ezp.bentley.edu/login?url=http://search.proquest.com/?url=http://search. proquest.com/docview/853194588?accountid=8576 (accessed 15 July 2013).

Barlett, Thomas. (2005). 'Course Examines Video Games as Cultural Indicators', 28 January. *Chronicle of Higher Education*, p. A10.

Barton, Ed. (2012). 'Global Advertising Spend to Increase by 4.9% in 2012 to Over $465 Billion', http://www.strategyanalytics.com/default.aspx?mod=press releaseviewer&a0=5180 (downloaded 13 June 2012).

Beaumont, Claudine. (2009). 'Apple's iPhone Is Proving to Be a Goldmine for Amateur Software Developers'. *Daily Telegraph*, 16 April, p. 22. http://proquest. umi.com/pqdweb?did=1679835551&Fmt=7&clientId=5258&RQT= 309&VName=PQD (accessed 15 July 2013).

Becker, Robert H. (1976). 'Future Is Now for Computer Games'. *Parks and Recreation* vol. 11 nos 12–13, p. 35.

Bell, Daniel. (1973). *The Coming of the Post-industrial Society*. New York: Basic Books.

Bergstein, Brian and AP. (2006a). 'IBM Cashes in on Chips: Cornering Market in Video Game Electronics Helps Bottom Line'. *Times Union*, p. C.1, http://proquest. umi.com/pqdweb?did=1435069811&Fmt=7&clientId=5258&RQT= 309&VName=PQD (accessed 15 July 2013).

Bergstein, Brian and AP. (2006b). 'Video Game War Galvanizes IBM's Chip-making Unit'. *St. Louis Post-Dispatch*, p. A.37. http://proquest.umi.com/pqdweb?did= 1160616551&Fmt=7&clientId=5258&RQT=309&VName=PQD (accessed 15 July 2013).

Bertz, Matt. (2010). 'Former Rockstar San Diego Employee Airs Dirty Laundry', 20 December, http://www.gameinformer.com/b/news/archive/2010/12/20/former-rockstar-san-diego-employee-airs-dirty-laundry.aspx (downloaded 10 October 2012).

Betts, M. (2009). 'Air Force Taps PlayStation 3 for Research'. *Computerworld* vol. 43 no. 34, p. 12.

Biersdorfer, J. D. (2004). 'Giving Gamers Another Window on Their World'. *New York Times*, 9 December, p. 280.

Bilton, Nick. (2011). 'Mobile App Revenue to Reach $38 Billion by 2015, Report Predicts', 28 February, http://bits.blogs.nytimes.com/2011/02/28/mobile-app-revenue-to-reach-38-billion-by-2015-report-predicts/ (downloaded 2 March 2011).

Bloom, David. (2001). 'Final Fantasy Goes to Hollywood'. *Red Herring*, 9 May, http://www.redherring.com/mag/issue96/1340019134.html (accessed 13 June 2001).

Bloom, David and Marc Graser. (2002). 'H'Wood Game to Cash in on Licenses'. *Variety*, 27 May–2 June, p. 10.

BLS. (2005a). 'Industry at a Glance', http://www.bls.gov/iag/iaghome.htm (accessed 12 May 2005).

BLS. (2005b). 'Industry at a Glance: Information', http://www.bls.gov/iag/iag. services.htm (accessed 12 May 2005).

Bogost, Ian. (2007). *Persuasive Games: The Expressive Power of Videogames*. Cambridge, MA: MIT Press.

Bogost, Ian, Simon Ferrari and Bobby Schweizer. (2010). *Newsgames: Journalism at Play*. Cambridge, MA: MIT Press.

Bolter, Jay David and Richard Grusin. (2000). *Remediation*. Cambridge, MA: MIT Press.

Bond, Paul. (2008). 'Video Game Sales on Winning Streak, Study Projects', http://www.reuters.com/article/technologyNews/idUSN1840038320080618 (downloaded 1 July 2008).

Bowen, Matt. (2005). 'Playdium.net Opens First U.S. Arcade'. *USA Today*, 12 April, http://usatoday30.usatoday.com/tech/news/2005-04-12-playdium-opens_x.htm (accessed 13 August 2013).

Box Office Mojo. (2011a). 'Video Game Adaptation, 1980–Present', http://boxofficemojo.com/yearly/ (downloaded 11 December 2011).

Box Office Mojo. (2011b). Yearly Box Office, http://boxofficemojo.com/yearly/ (downloaded 11 December 2011).

Branch, Taylor. (2011). 'The Shame of College Sports'. *Atlantic Monthly* vol. 308, pp. 80–4, 86, 88–9, 93–4, 96, 98, 100–102, 104, 106, 108, 110, http://ezp.bentley.edu/login?url=http://search.proquest.com/docview/898362622?accountid=8576 (accessed 15 July 2013).

Brandt, Richard. (1987). 'Trip Hawkins Wants to Be the Walt Disney of Software'. *Business Week*, 9 November, p. 134.

Brandt, Richard. (1990). 'Creeping up on the Mario Brothers'. *Business Week*, 30 July, pp. 74, 76.

Breznican, Anthony. (2004). ' "San Andreas" Hijacks Game Awards'. *Associated Press*, 15 December, http://apnews.excite.com/article/20041215/D8702IKG0.html (downloaded 19 December 2004).

Brickner, Sara. (2004). 'Going to War with A Stick'. *Eugene Weekly*, 2005, 16 December, http://www.eugeneweekly.com/2004/12/16/lastminutegifts.html (downloaded 19 December 2004).

Brodesser, Claude. (2005). 'Gamers' SAG Snag: Deadline Looms for Producers Unions'. *Variety*, 13 April, p. 1.

Brodesser, Claude and Ben Fritz. (2005). 'Halo, Hollywood'. *Variety*, 3 February, p. 1.

Brodesser, Claude and Dave McNary. (2005). 'Vidgame Biz and Thesps Extend Pact a Third Time'. *Variety*, 17 April, p. 2.

Brookey, Robert Alan. (2010). *Hollywood Gamers: Digital Convergence in the Film and Video Game Industries*. Bloomington: Indiana University Press.

Bruno, Antony. (2009). 'Gaming the System'. *Billboard* vol. 121 no. 3, p. 12.

Bruno, Antony and Mitchell Peters. (2009). 'Games Beatles Play'. *Billboard* vol. 121 no. 36, pp. 22–3.

Bulik, Beth Snyder. (2004). 'Hot Shot Marketing'. *Advertising Age* no. 75, 24 May, p. S-1.

Bulik, Beth Snyder. (2010). 'iPad Out to Prove Itself as Gaming Platform, but Will Users Play Along?' *Advertising Age* vol. 81 no. 13, p. 2.

Bulkeley, William M. (2003). 'Sony Plays a Videogame Grid'. *Wall Street Journal*, 27 February, p. B5.

Burt, Jeffrey. (2005). 'AMD, Intel Extend Manufacturing in Asia'. *eWeek*, vol. 22 no. 49. p. 17-17.

Business and Company Resource Center. (2011a). 'Company Profile: GameStop Corp'. *Business and Company Resource Center*, http://galenet.galegroup.com.ezp. bentley.edu/servlet/BCRC?rsic=PK&rcp=CO&vrsn=unknown&locID=bentley_main&srchtp=cmp&cc=1&c=1&mode=c&ste=60&tab=1&tbst=tsCM&ccmp=GameStop+Corp.&tcp=gamestop&docNum=DC1194259&bConts=3 (accessed 13 March 2012).

Business and Company Resource Center. (2011b). 'Company Profile: Valve L.L.C.'. *Business and Company Resource Center*, http://galenet.galegroup.com.ezp. bentley.edu/servlet/BCRC?rsic=PK&rcp=CO&vrsn=unknown&locID=bentley_main&srchtp=cmp&cc=29&c=29&mode=c&ste=60&tab=1&tbst=tsCM&ccmp=Valve+L.L.C.&n=25&tcp=valve&docNum=DC670528&bConts=3 (accessed 13 March 2012).

Business Week. (1976a). 'Atari Sells Itself to Survive Success'. *Business Week*, 15 November, pp. 120–1.

Business Week. (1976b). 'Demand Overwhelms Video Game Makers'. *Business Week*, 22 March, p. 31.

Business Week. (1976c). 'Colliding in a Low-price Market'. *Business Week*, 22 March, p. 63.

Business Week. (1979a). 'Hot Market in Electronic Toys'. *Business Week*, 17 December, pp. 108–10.

Business Week. (1979b). 'Why Electronic Games Will Be Hard to Find'. *Business Week*, 19 November, p. 52.

Business Week. (1985). 'How a Computer-game Maker Finesses the Software Slowdown'. *Business Week*, 10 June, p. 72.

Business Week. (1990). 'The Next Step up from Nintendo'. *Business Week*, 28 May, p. 107.

Business Wire. (2004). 'Atari Ships Unreal Tournament 2004: Latest in the Unreal Tournament Franchise Redefines PC Multiplayer Mayhem'. *Business Wire*, http://www.businesswire.com/news/home/20040316005575/en/Atari-Ships-Unreal-Tournament-2004-Latest-Unreal (accessed 13 August 2013).

Business Wire. (2009). 'Research and Markets: Latest Report Forecasts That Apple Will Help Drive Mobile Game Market Sales to $11.7 Billion by 2014'. *Business Wire*, http://www.businesswire.com/news/home/20091117006013/en/Research-Markets-Latest-Report-Forecasts-Apple-Drive (accessed 13 August 2013).

Business Wire. (2012a). 'EA SPORTS FIFA Soccer 13 Day One Sales up 42 Percent in North America'. *Business Wire*, http://ezp.bentley.edu/login?url=http://search. proquest.com/docview/1080797048?accountid=8576 (downloaded 17 October 2012).

Business Wire. (2012b). 'EA SPORTS FIFA Soccer 13 Poised for Historic Opening as Game Launches in North America'. *Business Wire*, http://ezp.bentley.edu/login?url=http://search.proquest.com/docview/1073475885?accountid=8576 (downloaded 17 October 2012).

Business Wire. (2012c). 'GameStop and Prima Games Announce "Buy in-Store, Redeem Online" eGuide Program'. *Business Wire*, http://ezp.bentley.edu/login?url=http://search.proquest.com/docview/1035271915?accountid=8576 (downloaded 17 October 2012).

Business Wire. (2012d). 'Madden NFL 13 Continues Historic Launch; Record Sell-through and Fan Engagement Highlight Week One'. *Business Wire*, http://ezp.bentley.edu/login?url=http://search.proquest.com/docview/1038139890?accountid=8576 (downloaded 17 October 2012).

Byster, Leslie A. and Ted Smith. (2006). 'The Electronics Production Life Cycle. From Toxics to Sustainability: Getting off the Toxic Treadmill'. In Ted Smith, David A. Sonnenfeld and David Naguib Pellow (eds). *Challenging the Chip: Labor Rights and Environmental Justice in the Global Electronics Industry*. Philadelphia, PA: Temple University Press, pp. 205–14.

Caldwell, Patrick. (2006). *Report: Korean Game Market to Surpass $2 Billion in 2007*. GameSpot, p. 285, Caldwell.

Calgary Herald. (2010). 'Xbox Eclipses $1B in Online Revenue'. *Calgary Herald*, p. E.3. http://proquest.umi.com/pqdweb?did=1814442741&Fmt=7&clientId=5258&RQT=309&VName=PQD (accessed 15 July 2013).

Campbell-Kelly, Martin. (2003). *From Airline Reservations to Sonic the Hedgehog: A History of the Software Industry*. Boston, MA: MIT Press.

Carillo, Maria Rosaria and Alberto Zazzaro. (2000). 'Innovation, Human Capital Destruction and Firms' Investment in Training'. *Manchester School* vol. 68 no. 3, pp. 331–48.

Carlson, Scott. (2003a). 'Southern Methodist U. Seeks to Train Game Designers'. *Chronicle of Higher Education*, 4 April.

Carlson, Scott. (2003b). 'Video Games Can Be Helpful to College Students, a Study Concludes'. *Chronicle of Higher Education*, 15 August, http://chronicle.com/article/Video-Games-Can-Be-Helpful-to/18156 (accessed 13 August 2013).

Carter, Bill. (2012). 'Prime-time Ratings Bring Speculation of a Shift in Habits'. *New York Times*, http://www.nytimes.com/2012/04/23/business/media/tv-viewers-are-missing-in-action.html?pagewanted=all (accessed 15 July 2013).

Castronova, Edward. (2001). *Virtual Worlds: A First-hand Account of Market and Society on the Cyberian Frontier* (Working Paper No. CESifo Working Paper No. 618). Munich: Centre for Economic Study & Ifo Institute for Economic Research.

Castronova, Edward. (2002). *On Virtual Economies* (Working Paper No. CESifo Working Paper No. 752). Munich: Centre for Economic Studies & Ifo Institute for Economic Research.

Castronova, Edward. (2005). *Synthetic Worlds: The Business and Culture of Online Games*. Chicago, IL: University of Chicago Press.

CERO. (2008). 'Ratings Explanation' (translated), http://translate.google.com/translate?u=http%3A%2F%2Fwww.cero.gr.jp%2Findex.html&hl=en&ie=UTF8&sl=ja&tl=en (downloaded 8 July 2008).

Cerrati, Michael. (2006). 'Video Game Music: Where It Came From, How It Is Being Used Today, and Where It Is Heading Tomorrow'. *Vanderbilt Journal of Entertainment and Technology Law* vol. 8 no. 2, pp. 293–334.

Chalk, Andy. (2010a). 'Rockstar Responds to San Diego Allegations', 15 January, http://www.escapistmagazine.com/news/view/97503-Rockstar-Responds-to-San-Diego-Allegations (downloaded 10 October 2012).

Chalk, Andy. (2010b). '"Rockstar Wives" Complain about Working Conditions', 11 January, http://www.escapistmagazine.com/news/view/97391-Rockstar-Wives-Complain-About-Working-Conditions (downloaded 10 October 2012).

Chazan, Guy. (2005). 'In Russia, Politicans Protect Movie and Music Pirates'. *Wall Street Journal*, 12 May, p. B.1.

Chiang, Oliver. (2011). 'The Master of Online Mayhem'. *Forbes* vol. 187 (no issue), 28 February pp. 30–2.

China Economic Review. (2010). 'NetEase Q3 Profit up 49% on Higher Games Sales'. *China Economic Review – Daily Briefings*, no issue, http://ezp.bentley.edu/login?url=http://search.proquest.com/docview/867349016?accountid=8576 (downloaded 3 March 2013).

China Economic Review. (2012). 'China's NetEase Receives Approval to Launch Game'. *China Economic Review – Daily Briefings*, http://ezp.bentley.edu/login?url=http://search.proquest.com/docview/1082275241?accountid=8576 (downloaded 3 March 2013).

Cho, Young-Sam. (2006). 'Chip Makers Can't Keep up with Demand around the Markets Marketplace by Bloomberg'. *International Herald Tribune*, p. 19, http://proquest.umi.com/pqdweb?did=1162849881&Fmt=7&clientId=5258&RQT=309&VName=PQD (accessed 15 July 2013).

Chung, Peichi and Anthony Fung. (2013). 'Internet Development and the Commercialization of Online Gaming in China'. In Nina B. Huntemann and Ben Aslinger (eds). *Gaming Globally: Production, Play, and Place*. New York: Palgrave Macmillan, pp. 233–50.

Citron, Alan and Leslie Helm. (1991). 'Sony Reported Near a Stock Offer Financing: Selling Shares Would Help Ease Debt from Acquiring CBS Records and Columbia Pictures'. *Los Angeles Times (pre-1997 Fulltext)*, p. 4, http://proquest.umi.com/

pqdweb?did=61013072&Fmt=7&clientId=5258&RQT=309&VName=PQD (accessed 15 July 2013).

Clabaugh, Jeff. (2010). 'HDTV Penetration Reaches 65 Percent', 6 May, http://www.bizjournals.com/washington/stories/2010/05/03/daily61.html (downloaded 10 October 2010).

CNN. (2003). 'Norway MP Plays PDA Game during War Debate', 31 January, http://www.cnn.com/2003/WORLD/europe/01/30/offbeat.norway.game.ap/index.html (downloaded 3 February 2003).

Cohen, David S. (2005). 'Record Revs but Lower Profits at Vidgamer EA: Net Income down from $577 Mil to $504 Mil in 2004', *Variety*, 3 May, http://www.google.com/url?sa=t&rct=j&q=&esrc=s&source=web&cd=1&ved=0CC4QFjAA&url=http%3A%2F%2Fwww.variety.com%2Farticle%2FVR1117922074%2F&ei=xNgLUu21BoHW2QXJzICgDQ&usg=AFQjCNFdzi83plbEf7V57xQ6xGQ_Et56wg&bvm=bv.50723672,d.b2I (accessed 13 August 2013).

Cohen, Scott. (1984). *Zap!: The Rise and Fall of Atari*. New York: McGraw-Hill.

Comas, Martin E. (2012). 'Candidates All Want to Focus on Attracting New Business'. *Orlando Sentinel*, p. J2.

Computer Weekly News. (2009). 'DFC Intelligence Forecasts That Apple Will Help Drive Mobile Game Market Sales to $11.7 Billion by 2014'. *Computer Weekly News*, p. 225.

Conditt, J. (2012). 'THQ Adaptation of Pixar's "Brave" Pops up in Australian Ratings'. *Joystiq.com*, http://www.joystiq.com/2012/01/19/thq-adaptation-of-pixars-brave-pops-up-in-australian-ratings/ (accessed 15 July 2013).

Condry, J. C. (1984). 'Mind and Media – the Effects of Television, Video Games and Computers'. *Smithsonian* vol. 5 no. 4, pp. 124–5.

Cook, John. (2006). 'A Moment with: CEO David Roberts, the "Adult Supervision" at PopCap Games'. *Seattle Post-Intelligencer*, 1 February, http://www.seattlepi.com/business/article/A-moment-with-CEO-David-Roberts-the-adult-1194466.php (accessed 13 August 2013).

Cooper, Gary L. and Eric K. Brown. (2002). *The Video Game Plan* (Research Brief). San Francisco, CA: Banc of America Securities, Equity Research.

Corcoran, Elizabeth. (2001). 'Promises, Promises'. *Forbes*, 19 February, pp. 54–6.

Coutts, Matthew. (2009). '"Play" Money; Real Dollars Can Be Made in the World of Virtual Gaming'. *National Post*, http://ezp.bentley.edu/login?url=http://search.proquest.com/docview/330927675?accountid=8576 (accessed 15 July 2013).

Croal, Ngai and Kay Itoi. (2004). 'Fall of the Video King'. *Newsweek* vol. 144 no. 18, October, pp. E30–E31.

Cuneo, Alice Z. (2002). 'Video-game Makers Clamor to Stand Out in Holiday Rush'. *Advertising Age* vol. 73, 28 October, p. 4.

Cuneo, Alice Z. (2004). 'Frank Gibeau'. *Advertising Age* vol. 75, 1 March, p. S-10.

Daily Gleaner. (2005). 'Xbox Unveils Processor'. *Daily Gleaner*, http://proquest.umi. com/pqdweb?did=983846701&Fmt=7&clientId=5258&RQT=309&VName= PQD (accessed 15 July 2013).

Dave and Buster's. (2005). 'Dave and Buster's History', http://www. daveandbusters.com//About/History.asps (downloaded 1 October 2005).

Davis, Avram. (2009). 'Disney Acquires Marvel Entertainment'. *Mergers & Acquisitions Report* vol. 22 no. 36, p. 13–13.

Dawtrey, Adam. (2005). 'U Has an Int'l Game Play'. *Variety* vol. 287 no. 60, 24 June, p. 4.

De La Merced, Michael J. (2011). 'Electronic Arts to Buy PopCap Games'. *New York Times*, http://dealbook.nytimes.com/2011/07/12/electronic-arts-to-buy-popcap-games/?_r=0 (accessed 13 August 2013).

Deagon, Brian. (2010). 'Real Dollars in Online Games' Virtual Goods 0 to 5 Mil Users in 1 Month $1 Each to Sell Items like Weapons to Most-devoted Game Players Is Adding Up'. *Investor's Business Daily*, 9 August, p. A04.

Dee, Jonathan. (2005). 'PlayStations of the Cross'. *New York Times Magazine*, 1 May, http://www.nytimes.com/2005/05/01/magazine/01GAMES.html (downloaded 3 May 2005).

Delaney, Kevin J. (2003). 'A Game Maker's Big Gamble'. *Wall Street Journal*, 14 May, p. B1.

DeMaria, Rusel. (2003). *High Score: The Illustrated History of Electronic Games*, 2nd edn. New York: McGraw-Hill Osbourne Media.

Denison, D. C. (2011). 'Harmonix Confirms Layoffs amid Sales Slump'. *Boston Globe*, p. B.5, http://ezp.bentley.edu/login?url=http://search.proquest.com/ docview/851206078?accountid=8576 (accessed 15 July 2013).

Deutsch, Claudia H. (2002). 'Game-design Courses Gain Favor'. *New York Times*, 1 April, http://www.nytimes.com/2002/04/01/technology/01GAME.html (accessed 15 July 2013).

Diamante, Vincent. (2005). 'E3 Report: New Practices in Licensing and Ancillary Rights', *Gamasutra*, 23 May, http://www.gamasutra.com/features/20050523/ diamante_pfv.html (downloaded 26 May 2005).

Disney. (2007a). *Annual Report, 2007*. Burbank, CA: Walt Disney Company, Inc.

Disney. (2007b). Company Information, http://disney.go.com/disneyinteractivestudios/ company.html (downloaded 22 July 2008).

Disney. (2007c). Board of Directors, http://corporate.disney.go.com/corporate/ board_of_directors.html (downloaded 22 July 2008).

Disney. (2012). 'Disney Interactive Studios – Company Information,' http://disney. go.com/disneyinteractivestudios/company.html (downloaded 8 August 2012).

DOC. (1997). '1997 Economic Census: Information, United States', http://www.census.gov/epcd/ec97/us//US000_51.HTM – N51 (downloaded 25 May 2002).

DOE. (2003). 'Video Games – Did They Begin at Brookhaven?', http://www.osti.gov/accomplishments/videogame.html (downloaded 20 January 2002).

DOL. (2005). *2004–2005 Occupational Outlook Handbook*. Washington, DC: US Department of Labor.

Donaton, Scott. (2003). 'We Unravel the Mystery of Where the Boys (18–34) Are'. *Advertising Age* vol. 74, 3 November, p. 23.

Donaton, Scott. (2005). 'Can Video-game Characters Sing? MTV2 Has the Answer'. *Advertising Age* vol. 76, 3 January, p. 14.

Donovan, Tristan. (2004). 'Top 20 Publishers'. *Game Developer* vol. 11, pp. 9–15.

Douglas, Christopher. (2002). '"You Have Unleashed a Horde of Barbarians": Fighting Indians, Playing Games, Forming Disciplines'. *Postmodern Culture* vol. 13 no. 1, http://muse.jhu.edu/ (accessed 13 August 2013).

Duffy, Jill. (2007a). 'Game Art and Animation, an Introduction', 20 August, http://www.gamecareerguide.com/features/413/features/413/game_art_and_animation_an_.php (downloaded 10 October 2012).

Duffy, Jill. (2007b). 'Game Production, an Introduction', 20 August, http://www.gamecareerguide.com/features/414/game_production_an_introduction.php (downloaded 10 October 2012).

Duffy, Jill. (2007c). 'Game Programming, an Introduction', 20 August, http://www.gamecareerguide.com/features/412/features/412/game_programming_an_.php (downloaded 10 October 2012).

Dyer-Witheford, Nick. (1999). 'The Work in Digital Play: Video Gaming's Transnational and Gendered Division of Labor'. *Journal of International Communication* vol. 6 no. 1, pp. 69–93.

Dyer-Witheford, Nick. (2002). 'Cognitive Capital Contested: The Class Composition of the Video and Computer Game Industry', November, http://multitudes.samizdat.net/article.php?id_article=268 (accessed 15 July 2013).

Dyer-Witheford, Nick and Greig De Peuter. (2009). *Games of Empire: Global Capitalism and Video Games*. Minneapolis: University of Minnesota Press.

Dyer-Witheford, Nick and Zena Sharman. (2005). 'The Political Economy of Canada's Video and Computer Game Industry'. *Canadian Journal of Communication* vol. 30 no. 2, pp. 187–210.

EA. (2005). *Annual Report, 2005*. Redwood City, CA: Electronic Arts, Inc.

EA. (2006). *Annual Report, 2006*. Redwood City, CA: Electronic Arts, Inc.

EA. (2007). *Annual Report, 2007*. Redwood City, CA: Electronic Arts, Inc..

EA. (2008). *Annual Report, 2008*. Redwood City, CA: Electronic Arts, Inc.

EA. (2010). Company Overview – Locations, http://aboutus.ea.com/locations.action (downloaded 11 May 2010).

ea_spouse. (2004a). 'EA: The Human Story', 20 November, http://www.livejournal.
   com/users/ea_spouse/274.html (downloaded 21 January 2005).

ea_spouse. (2004b). 'Followup to EA: The Human Story', December,
   http://www.livejournal.com/users/ea_spouse/274.html (downloaded 21 January
   2005).

Economist. (2004). 'Deus ex machinima?'. *Economist*, 16 September,
   http://www.economist.com/science/tq/PrinterFriendly.cfn?Story_ID=3171417
   (accessed 13 August 2013).

Economist. (2005). 'Gaming's Next Level; Rational Consumer'. *Economist* vol. 376
   no. 8444, p. 18.

Economist. (2009). 'Tech View: Of Dragons and Dungeons'. *Economist.com*,
   http://www.economist.com/node/13635572 (accessed 13 August 2013).

Economist. (2011). 'Jedi v Orc'. *Economist (Online)*, http://ezp.bentley.edu/
   login?url=http://search.proquest.com/docview/911989538?accountid=8576
   (accessed 15 July 2013).

Edwards, Cliff. (2012). 'Activision, the Anti-Zynga'. *Business Week*,
   http://www.businessweek.com/articles/2012-10-04/activision-the-anti-zynga
   (accessed 15 July 2013).

Elkin, Tobi. (2002a). 'Bewitched by Games'. *Advertising Age* vol. 73, 10 June, p. 62.

Elkin, Tobi. (2002b). 'Sony Ties MD Walkman to Online Game'. *Advertising Age*
   vol. 73, 10 June, p. 62.

Elkin, Tobi. (2002c). 'Video Games Try Product Placement'. *Advertising Age* vol. 73,
   20 May, p. 157.

Elkin, Tobi. (2002d). 'Video-game Makers Heat up Competition'. *Advertising Age*
   vol. 73, 20 May, pp. 3, 157.

Elkin, Tobi. (2003). 'Struggling Toy Industry Looks to Licensing'. *Advertising Age*,
   vol. 74, 17 February, pp. 4, 36.

ELSPA. (2004). *Chicks and Joysticks: An Exploration of Women and Gaming*. London:
   Entertainment and Leisure Software Publishers Association.

Emeling, Shelley. (2004). 'Seniors Taking to Computer Games'. *Salt Lake Tribune*,
   30 December.

Entertainment and Business Newsweekly. (2010). 'Business News: Is Digital
   Distribution Reshaping the Gaming Industry? Answer: Not Necessarily'.
   Entertainment Business Newsweekly, p. 115.

Entertainment and Business Newsweekly. (2012). 'Electronic Arts Inc.: EA Sports
   NHL 13 Skates to New Franchise High in First Week'. *Entertainment Business
   Newsweekly*, p. 375.

Entertainment Newsweekly. (2009). 'Research and Markets: New Mobile Games
   Platforms: The Challenges and Opportunities Facing This Industry'. *Entertainment
   Newsweekly*, p. 166.

Entertainment Newsweekly. (2012). 'Electronic Arts Inc.; EA SPORTS FIFA Soccer 13 Day One Sales up 42 Percent in North America'. *Entertainment Newsweekly*, p. 313.

Erard, Michael. (2004). 'In These Games, the Points Are All Political'. *New York Times*, p. G.1, http://proquest.umi.com/pqdweb?did=657448521&Fmt= 7&clientId=5258&RQT=309&VName=PQD (accessed 15 July 2013).

ESA. (2001). *Video Games and Youth Violence: Examining the Facts*. Washington, DC: Entertainment Software Association.

ESA. (2002). *Who Plays Computer and Video Games*. Entertainment Software Association, http://www.idsa.com/ffbox2.html (downloaded 14 February 2002).

ESA. (2004). *2004 – Sales Demographic, and Usage Data: Essential Facts about the Computer and Video Game Industry*. Entertainment Software Association.

ESA. (2005a). *2005 Sales Demographic, and Usage Data. Essential Facts about the Computer and Video Game Industry*. Entertainment Software Association.

ESA. (2005b). *Top Ten Industry Facts*. Entertainment Software Association, http://www.theesa.com/facts/top_10_facts.php (downloaded 10 March 2005).

ESA. (2006). *2006 Sales, Demographic, and Usage Data: Essential Facts about the Computer and Video Game Industry*. Entertainment Software Association.

ESA. (2007). *2007 Sales, Demographic, and Usage Data: Essential Facts about the Computer and Video Game Industry*. Entertainment Software Association.

ESA. (2008). *2008 Sales, Demographic, and Usage Data: Essential Facts about the Computer and Video Game Industry*. Entertainment Software Association.

ESA. (2009). *2009 Sales, Demographic, and Usage Data: Essential Facts about the Computer and Video Game Industry*. Entertainment Software Association.

ESA. (2010). *2010 Sales, Demographic, and Usage Data: Essential Facts about the Computer and Video Game Industry*. Entertainment Software Association.

ESA. (2011). *2011 Sales, Demographic, and Usage Data: Essential Facts about the Computer and Video Game Industry*. Entertainment Software Association.

ESA. (2012). *2012 Sales, Demographic, and Usage Data: Essential Facts about the Computer and Video Game Industry*. Entertainment Software Association.

ESA Canada. (2012). *Essential Facts 2012*. Entertainment Software Association.

ESRA. (2010). External Game Rating: Foreigner Game, http://ircg.ir/index.php?sn= foreignerGames&pt=list&&lang=en (downloaded 11 May 2010).

Euromonitor International. (2005). *Executive Summary: Toys and Games in China*. World Wide Web: Euromonitor International.

Fahey, Rob. (2003). 'Sega Sees Red over Simpsons Road Rage'. Gameindustry.biz, 13 May, http://www.gamesindustry.biz/articles/sega-sees-red-over-simpsons-road-rage (accessed 23 January 2004).

Farhi, Paul. (1994). 'Sony's Surprise: It Says It Overpaid for Columbia'. *Washington Post*, p. B.01, http://proquest.umi.com/pqdweb?did=72296774&Fmt= 7&clientId=5258&RQT=309&VName=PQD (accessed 15 July 2013).

Felberbaum, Michael. (2005). 'Nintendo Mobile Game Machine Goes Online'. *Associated Press*, 10 May, http://apnews.excite.com/article/20050510/ D8A0CN0o00.html (downloaded 22 May 2005).

Ferris, Timothy. (1977). 'Solid-state Fun'. *Esquire* vol. 87, March, pp. 100–1, 121.

Fidler, Roger. (1997). *Mediamorphosis*. Thousand Oaks, CA: Pine Forge Press.

FinancialWire. (2007). 'Video Game Console Wars Benefit Chip Makers'. *FinancialWire*, p. 1.

Fincher, Jack. (1978). 'Those Home Video Games Are Humming!'. *Reader's Digest* vol. 112, March, pp. 157–60.

Flare. (2012). 'Social Media Earning to Hit $16.9bn'. *Flare*, p. 36–36, http://ezp. bentley.edu/login?url=http://search.proquest.com/docview/1095686483?accountid =8576 (accessed 15 July 2013).

Florida, Richard. (2002). *The Rise of the Creative Class*. New York: Basic Books.

Flynn, Laurel J. (2005). 'Electronic Arts Lowers Forecast, Citing Machine Shortages'. *New York Times*, 22 March, http://www.nytimes.com/2005/03/22/technology/ 22art.html (accessed 13 August 2013).

Forbes. (1977). 'The Good Life Beckons', *Forbes*, 15 April, p. 53.

Forbes, Peter. (2011). 'The Heart of the Matter'. *Independent*, p. 26, http://ezp.bentley. edu/login?url=http://search.proquest.com/docview/852582516?accountid=8576 (accessed 15 July 2013).

Fordahl, Matthew. (2005). 'Sony Fined $90.7M in PlayStation Case'. *Associated Press*, 28 March, http://www.washingtonpost.com/wp-dyn/articles/A7785- 2005Mar28_2.html (accessed 13 August 2013).

Foster, Andrea L. (2004a). 'A Scholar Who Brings Philosophy to Video Games'. *Chronicle of Higher Education*, 29 October, p. A33.

Foster, Andrea L. (2004b). 'Video Games with a Political Message'. *Chronicle of Higher Education*, 29 October, pp. A32–33.

Frith, Simon. (1997). 'Entertainment'. In James Curran and Michael Gurevitch (eds). *Mass Media and Society*, 2nd edn. London: Arnold, pp. 160–76.

Fritz, Ben. (2004). 'Hot Vidgames Snub H'wood'. *Variety*, 21 December, p. 1.

Fritz, Ben. (2005a). 'ESPN Playing Ball with EA for Videogames'. *Variety*, 17 January, p. 6.

Fritz, Ben. (2005b). 'Less Game EA Cuts Jobs at L.A. Facility'. *Variety*, 27 January, http://variety.com/2005/digital/news/less-game-ea-cuts-jobs-at-l-a-facility-2- 1117917028/ (accessed 13 August 2013).

Fritz, Ben. (2005c). 'Mouse Hunts Vidgamers: Disney Falls for Avalanche'. *Variety*, 19 April, p. 4.

Fritz, Ben. (2005d). 'Pirates' Plys 'Net: Disney Plans Multiplayer Online Game Tie-in'. *Variety*, 20 April, p. 3.

Fritz, Ben. (2005e). 'Power Rangers Gobble up Pac-Man: Toy Co. Buys up Videogame Giant Namco for $1.7 Bil'. *Variety*, 2 May, p. 2.

Fritz, Ben. (2005f). 'SAG Rejects Vidgame Deal'. *Variety*, 21 June, p. 1.

Fritz, Ben. (2005g). 'SAG Revisits Vidgame Agreement'. *Variety*, 23 June, p. 5.

Fritz, Ben. (2005h). 'Sony Plays 'Matrix': WB Sells off Rights to Online Videogame'. *Variety*, 16 June, p. 6.

Fritz, Ben. (2005i). 'Sony Takes Page from Ubisoft's Game Plan'. *Variety*, 26 January, p. 8.

Gaar, Brian. (2011). 'Video Game Maker Loses Job, Lays off 39 Workers'. *Austin American Statesman*, http://ezp.bentley.edu/login?url=http://search.proquest.com/docview/902399480?accountid=8576%3E) (accessed 15 July 2013).

Gamasutra. (2005a). 'Electronic Arts Signs Exclusive Deal with NCAA'. *Gamasutra.com*, 11 April, http://www.gamasutra.com/view/news/96229/Electronic_Arts_Signs_Exclusive_Deal_With_NCAA.php (downloaded 12 August 2013).

Gamasutra. (2005b). 'European PlayStation 2 Users Base to Exceed U.S'. *Gamasutra.com*, 7 April, http://www.gamasutra.com/view/news/96218/European_PlayStation_2_User_Base_To_Exceed_US.php (downloaded 12 August 2013).

Gamasutra. (2005c). 'GameStop Sees Sales Growth, Profits Down'. *Gamasutra.com*, 23 March, http://www.gamasutra.com/view/news/5173/GameStop_Sees_Sales_Growth_Profits_Down.php (downloaded 23 March 2005).

Gamasutra. (2005d). 'Survey Shows Top Game Businesses Make $25 Billion'. *Gamasutra.com*, 22 March, http://www.gamasutra.com/view/news/5166/c (downloaded 23 March 2005).

Gamasutra. (2007). 'Mario Tops Best Selling Franchise List'. *Gamasutra.com*, 10 January, http://www.gamasutra.com/php-bin/news_index.php?story=12349 (downloaded 18 July 2007).

Gamasutra. (2008). 'NPD: 2007 U.S. Game Industry Growth up 43% to $17.9 Billion'. *Gamasutra.com*, 18 January, http://gamasutra.com/php-bin/news_index.php?story=17006 (downloaded 1 July 2008).

Gamasutra. (2009). 'GDC Austin: An Inside Look at the Universe of Warcraft'. *Gamaustra.com*, 17 September, http://www.gamasutra.com/view/news/25307/GDC_Austin_An_Inside_Look_at_the_Universe_Of_Warcraft.php (downloaded 11 May 2010).

Gamasutra. (2012). 'Game Developer Reveals 2011 Game Industry Salary Survey Results'. *Gamasutra.com*, 2 April, http://gamasutra.com/view/news/167355/Game_Developer_reveals_2011_Game_Industry_Salary_Survey_results.php -.UIihNYU8flo (downloaded 18 July 2012).

Game Career Guide. (2012). 'Game Career Guide – Schools', http://www.gamecareerguide.com/schools/ (downloaded 10 October 2012).

Game Politics.com. (2008). 'Report: Rare Metal Fueled African "PlayStation War"',
   11 July, http://www.gamepolitics.com/2008/07/11/report-rare-metal-fueled-african-
   quotplaystation-warquot (downloaded 18 August 2008).

Gamedaily.biz. (2005). 'US Console and PC Software Sales Set New High in 2004'.
   Gamedaily.biz, 31 January, http://www.majorleaguegaming.com/news/u-s-console-
   and-pc-software-sales-set-new-high-in-2004 (accessed 13 August 2013).

Gamedevmap. (2010). 'Game Developer Map', http://www.gamedevmap.com/
   (downloaded 11 May 2010).

Garity, Brian. (2003). 'Gaming Sales Rose 10% Last Year'. *Billboard* vol. 115 no. 8,
   February, p. 38.

Gaudiosi, John. (2012). 'New Reports Forecast Global Video Game Industry Will
   Reach $82 Billion by 2017', 18 July, http://www.forbes.com/sites/johngaudiosi/
   2012/07/18/new-reports-forecasts-global-video-game-industry-will-reach-82-billion-
   by-2017/ (downloaded 24 May 2012).

Gentile, Gary. (2005a). 'Actors Weigh Strike over Video Game Voices'. *USA Today*,
   25 May, http://usatoday30.usatoday.com/tech/news/2005-05-25-voice-actor-
   protest_x.htm (accessed 13 August 2013).

Gentile, Gary. (2005b). 'Products Placed Liberally in Video Games'. *Associated Press*,
   20 May, http://apnews.excite.com/article/20050521/D8A789200.html (accessed
   27 May 2005).

Gibbs, Colin. (2007). 'Online Gaming Giant Sets Sights on Mobile Space'. *RCR
   Wireless News*, vol. 26, 4 June, p. 18.

Gnatek, Tim. (2005). 'With a PC's Power, That's Entertainment'. *New York Times*,
   21 April, p. C.9.

Goetz, Thomas. (2012). 'How to Spot the Future'. *Wired*, vol. 20, http://ezp.bentley.
   edu/login?url=http://search.proquest.com/docview/1021206662?accountid=8576
   (accessed 15 July 2013).

Good, Owen. (2011). 'You Can Have Any Sports Video Game You Want, but There's
   Only One'. *Kotaku.com*, http://kotaku.com/5806547/you-can-have-any-sports-
   video-game-you-want-but-theres-only-one (downloaded 17 May 2012).

Graft, Kris. (2009). 'Sony Shrinks PS3 Build Costs by 70 Percent'. *Gamasutra.com*,
   31 July, http://www.gamasutra.com/php-bin/news_index.php?story=24659
   (downloaded 17 December 2010).

Graft, Kris. (2010a). 'Report: Wii Holds 47 Percent of Global Console Revenue'.
   Gamasutra, http://www.gamasutra.com/view/news/27932/Report_Wii_Holds_47_
   Percent_Of_Global_Console_Revenue.php (accessed 12 August 2013).

Graft, Kris. (2010b). 'Rockstar: "Typical" Layoffs Hit Red Dead Redemption Studio'.
   Gamasutra.com, 15 July, http://www.gamasutra.com/view/news/29457/Rockstar_
   Typical_Layoffs_Hit_Red_Dead_Redemption_Studio.php - .UI4MioU8flo
   (downloaded 10 October 2010).

Grant, Peter and Bruce Orwall. (2004). 'Disney Buys Henson's Muppets: Comcast Continues Its Pursuit'. *Wall Street Journal – Eastern Edition*, vol. 243 no. 33, p. A3.

Graser, Marc. (2009). 'Games Rock Song Sales'. *Variety*, vol. 416, p. 14.

Greenfield, Rebecca. (2013). 'Blacks and Latinos Aren't Thriving in Silicon Valley's Meritocracy', 7 February, http://www.theatlanticwire.com/technology/2013/02/blacks-and-latinos-arent-thriving-silicon-valleys-meritocracy/61890/ (downloaded 12 April 2013).

Greenpeace. (2007a). 'Clash of the Consoles: Battle for a Green Future', 11 December, http://www.greenpeace.org/international/en/news/features/clashoftheconsoles111207/ (downloaded 10 July 2010).

Greenpeace. (2007b). 'Nintendo, Microsoft and Phillips Flunk Toxic Test', 27 November, http://www.greenpeace.org/international/en/news/features/greener-electronics-ranking-6-291107/ (downloaded 10 July 2010).

Greenpeace. (2008). 'Game Consoles: No Consolation', 20 May, http://www.greenpeace.org/international/en/news/features/game-consoles-no-consolation200508/ (downloaded 10 July 2010).

Grossman, Lev and Evan Narcisse. (2011). 'Conflict of Interest'. *Time* vol. 178 no. 17, pp. 70–5.

Grover, Ronald. (1998). 'The Wonderful World of Disney.Com'. *Business Week*, 2 March, p. 78.

Grover, Ronald, Cliff Edwards, Ian Rowley and Moon Ihlwan. (2005). 'Game Wars: Who Will Win Your Entertainment Dollar, Hollywood or Silicon Valley?'. *Business Week*, 29 February, pp. 60–6.

Guth, Robert A. (2001). 'Sony Game Unit Invests in Square, Preparing for Fight with Microsoft'. *Wall Street Journal*, 10 October, p. A.13.

Guth, Robert A. (2002). 'PlayStation 2 Helps Sony Beat Forecasts'. *Wall Street Journal*, 28 January, p. A.12.

Guth, Robert A. (2003). 'Sega Tries New Game in Merger with Pachinko Maker Sammy – Software Designer's Pact Brings a Debt-free Partner with High Equity Returns'. *Wall Street Journal*, 14 February, p. B.5.

Guth, Robert A. (2005). 'Getting Xbox 360 to Market: Microsoft Must Coordinate Game Player's 1,700 Parts to Ensure Big Enough Supply'. *Wall Street Journal*, p. B.1.

Guth, Robert A. and Khan T. L. Tran. (2002). 'A Global Journal Report: Nintendo Goes Gory – Mario Bros.' Creator Courts Adult Gamers as Competition from Sony, Microsoft Grows'. *Wall Street Journal*, 14 October, p. B.1.

Guth, Robert A. and Phred Divorack. (2005). 'Microsoft, Sony Enter Epic Battle'. *Wall Street Journal*, 10 May, p. B.1.

Guth, Robert A., Nick Wingfield and Phred Divorack. (2005). 'It's Xbox 360 vs. PlayStation 3, and War Is About to Begin'. *Wall Street Journal*, 9 May, p. B.1.

Gutnick, Aviva Lucas, Michael Robb, Lori Takeuchi and Jennifer Kotler. (2011). *Always Connected: The New Digital Media Habits of Young Children*. New York: Joan Ganz Cooney Center at Sesame Workshop, p. 48.

Gwinn, Eric. (2009). 'Nintendo's New DSi Squares off with PSP, iPod Touch'. *McClatchy – Tribune News Service*.

Handrahan, Matthew. (2013). 'Sony's PlayStation Profits Slump 86 Per Cent in Q3', http://www.gamesindustry.biz/articles/2013-02-07-sonys-playstation-profits-slump-86-per-cent-in-q3?utm_source=newsletter&utm_medium=email&utm_campaign=us-daily (downloaded 23 March 2013).

Harmon, Amy. (1993). 'Sony Will Enter Video Game Field'. *LATimes.com*, 28 October, http://articles.latimes.com/1993-10-28/business/fi-50746_1_video-game-market (downloaded 23 March 2013).

Harmon, Amy. (1994). 'Redux of Atari Crash of '83? Video Game Machine Sales Are Expected to Be off 15%'. *Los Angeles Times*, p. 1. http://proquest.umi.com/pqdweb?did=6274933&Fmt=7&clientId=5258&RQT=309&VName=PQD. (downloaded 23 March 2013).

Harris, Dana. (2004). 'Boll Scores $47 Mil'. *Variety*, 4 January, p. 6.

Harris, Ron. (2005). ' "Walk of Game" Honors Video Game Icons'. *USA Today*, 9 March, http://usatoday30.usatoday.com/tech/news/2005-03-09-walk-of-game_x.htm (accessed 13 August 2013).

Hartley, Matt. (2010). 'A 1.5-pound Electronic Heavyweight Enters Ring'. *National Post*, p. A.1, http://proquest.umi.com/pqdweb?did=1969843741&Fmt=7&clientId=5258&RQT=309&VName=PQD (accessed 15 July 2013).

Hayes, Michael, Stuart Dinsey and Nick Parker. (1995). *Games War: Video Games – A Business Review*. London: Bowerdean Publishing Company Limited.

Haynes, Peter. (1994). 'The Computer Industry: The Third Age'. *Economist* no. 332, 17 September, p. SS3.

Health & Beauty Close-up. (2012). 'GameStop to Stock Prima eGuides'. *Health & Beauty Close-Up*, Jacksonville, Florida.

Hein, Kenneth. (2002). 'EA Games Star Potter, Bond'. *Brandweek* vol. 43, 28 October, p. 12.

Hemphill, Thomas A. (2002). 'Self-regulation, Public Issue Management and Marketing Practices in the U.S. Entertainment Industry'. *Journal of Public Affairs*, vol. 3 no. 4, pp. 338–57.

Herold, Charles. (2003a). 'Amid the Fighters, a Few Winners'. *New York Times*, 25 December, p. G.5.

Herold, Charles. (2003b). 'A Chase Saga Dominated by Its Cinematic Cousin'. *New York Times*, 15 May, p. G.7.

Herz, J. C. (1997). *Joystick Nation: How Videogames Ate Our Quarters, Won Our Hearts, and Rewired Our Minds*. Boston, MA: Little, Brown and Company.

Hewwit, Don. (2005). E-mail Interview. Washington, DC.

Hilbert, Martin and Priscila López. (2011). 'The World's Technological Capacity to Store, Communicate, and Compute Information'. *Science* vol. 332 no. 60, pp. 60–5.

Hillis, Scott. (2008). 'Nintendo Online Offerings Add Interest with WiiWare'. *Times – Colonist*, p. C.10, http://proquest.umi.com/pqdweb?did=1904315921&Fmt= 7&clientId=5258&RQT=309&VName=PQD (accessed 12 August 2013).

Ho, Chun Kit, Päivi Pöyhönen and Eeva Simola. (2009). *Playing with Labour Rights: Music Player and Game Console Manufacturing in China*. Helsinki: FinnWatch.

Holson, Laura M. (2005a). 'Blockbuster with a Joystick'. *New York Times*, 7 February, http://www.nytimes.com/2005/02/07/business/07game.html?_r=0 (accessed 13 August 2013).

Holson, Laura M. (2005b). 'A New Hollywood Player Pushes a Different Game'. *New York Times*, 16 May, p. C.1.

Hoover's. (2002). 'Electronic Arts – Capsule', http://www-2.hoovers.com/co/ capsule/9/0,2163,14059,00.html (downloaded 2 December 2002).

Hoover's. (2005). 'Hoovers Corporate Profiles', http://www.hoovers.com (downloaded 8 October 2005).

Hoover's. (2008). 'Hoovers Corporate Profiles', http://www.hoovers.com (downloaded 13 July 2008).

Hoover's. (2010). 'SEGA Corporation', http://www.hoovers.com (downloaded 3 August 2010).

Hoover's. (2012). 'Hoovers Corporate Profiles', http://www.hoovers.com (downloaded 17 May 2012).

Howells, Sacha A. (2002). 'Watching a Game, Playing a Movie: When Media Collide'. In Geoff King and Tanja Krzywinska (eds). *Screenplay: Cinema/Videogames/ Interfaces*. London: Wallflower Press, pp. 110–21

Huebner, Daniel. (2002). 'The Other Consoles'. *Game Developer* vol. 9 no. 2, p 6.

Huguenin, Patrick. (2009). 'A Big App-Etite. Gamers Find a New Playing Field with iPhone Downloads'. *New York Daily News*, p. 10. http://proquest.umi.com/ pqdweb?did=1849667851&Fmt=7&clientId=5258&RQT=309&VName=PQD (accessed 12 August 2013).

Huntemann, Nina B. (2009). 'Interview with Colonel Casey Wardynski'. In Nina B. Huntemann and Matthew Thomas Payne (eds). *Joystick Soldiers: The Politics of Play in Military Video Games*. New York: Routledge, pp. 178–88

Huntemann, Nina B. and Matthew Thomas Payne (eds). (2009). *Joystick Soldiers: The Politics of Play in Military Video Games*. New York: Routledge.

Huntemann, Nina. (2010). 'Irreconcilable Differences: Gender and Labor in the Video Game Workplace'. *FlowTV.org* vol. 11 no. 6, 22 January, http://flowtv.org/2010/01/irreconcilable-differences-gender-and-labor-in-the-video-game-workplace-nina-b-huntemann-suffolk-university/ (downloaded 10 October 2010).

Hutaff, Matt. (1996). 'Jilted, Lethargic Video-game Testers Unite!'. *Daily Trojan*, 21 October, http://matthutaff.com/college-clippings/1996/jilted-lethargic-video-game-testers-unite.html (accessed 13 August 2013).

IBISWorld. (2012). *DVD, Game & Video Rental*. Santa Monica, CA: IBISWorld.

IDSA. (2003). *Essential Facts about the Computer and Video Game Industry: 2003 Sales, Demographics and Usage Data*. Washington, DC: Interactive Digital Software Association.

IEAA. (2005). *Game Play Australia 2005*. Eveleigh: Interactive Entertainment Association of Australia, http://www.igea.net/wp-content/uploads/2009/08/GamePlay_Aust_2005.pdf (accessed 12 August 2013).

IEAA. (2007). *Interactive Australia 2007*. Eveleigh: Interactive Entertainment Association of Australia, http://www.igea.net/2007/01/interactive-australia-2007/ (accessed 12 August 2013).

IGDA. (2004a). *2004 Persistent Worlds White Paper*. Mount Royal, NJ: International Game Developer Association, http://www.igda.org/online/IGDA_PSW_Whitepaper_2004.pdf (accessed 12 August 2013).

IGDA. (2004b). *Quality of Life Survey*, Mount Royal, NJ: International Game Developers Association, http://www.igda.org/quality-life-white-paper-info (accessed 12 August 2013).

IGDA. (2005). Breaking in – Preparing for Your Career in Games: School Listings, http://www.igda.org/breakingin/resource_schools.php (downloaded 10 June 2005).

Ihlwan, Moon. (2007). 'South Korea: Video Games' Crazed Capital'. *Business Week*, 26 March, http://www.businessweek.com/globalbiz/content/mar2007/gb20070326_937184.htm (downloaded 2 July 2007).

Inside the Internet. (2000). 'Sega or Sony: Which Machine Will You Buy?'. *Inside the Internet* vol. 7 no. 1, p. 15.

International Herald Tribune. (2007). 'Sony to Sell Video Game 'Cell' Chip to Toshiba: Focus on Electronics Is Part of New Plan'. *International Herald Tribune*, p. 11, http://proquest.umi.com/pqdweb?did=1692002441&Fmt=7&clientId=5258&RQT=309&VName=PQD (accessed 15 July 2013).

IOC. (1999). *Olympic Marketing Matters*. Lausanne: International Olympic Committee.

IOC. (2000). *Olympic Marketing Matters*. Lausanne: International Olympic Committee.

IOC. (2008). *IOC Marketing Report – Beijing 2008*. Lausanne: International Olympic Committee.

IOC. (2010). *IOC Marketing Report – Vancouver 2010*. Lausanne: International Olympic Committee.

IOC. (2012a). *IOC Marketing Report – London 2012*. Lausanne, Switzerland: International Olympic Committee.

IOC. (2012b). *Olympic Marketing Fact File, 2012 Edition*. Lausanne, Switzerland: International Olympic Committee.

Ip, Barry, and Gabriel Jacobs. (2004). 'Territorial Lockout: An International Issue in the Videogames Industry'. *European Business Review*, vol. 16 no. 5, pp. 511–21.

Irish Times. (2010). 'Xbox Live Online Exceeds Estimates with [Euro]1bn in Sales'. *Irish Times*, p. 8. http://proquest.umi.com/pqdweb?did=1543339741&Fmt=7&clientId=5258&RQT=309&VName=PQD (accessed 15 July 2013).

Irvine, Martha. (2004). 'Web Gaming Changes Social Interactions'. *Associated Press*, 5 December, http://apnews.excite.com/article/20041206/D86ptpu00.html (downloaded 13 January 2005).

Itzkoff, Dave. (2008). 'Obama Ads Appear in Video Game'. *New York Times*, p. C.2. http://ezp.bentley.edu/login?url=http://search.proquest.com/docview/433955009?accountid=8576 (accessed 15 July 2013).

Ivry, Sara. (2005). 'Best Buy to Toughen Policy on Adult Game Sales'. *New York Times*, 23 May, http://www.nytimes.com/2005/05/23/technology/23game.html (accessed 13 August 2013).

Iwabuchi, Koichi. (2002). *Recentering Globalization: Popular Culture and Japanese Transnationalism*. Durham, NC: Duke University Press.

Jamie Kirschenbaum, Mark West, and Eric Kearns v. *Electronic Arts, Inc* (2004), p. 16.

Johnson, Bobbie. (2007). 'Halo 3 Enters the Fray in £19bn Video Games Battle', 26 September, http://www.guardian.co.uk/technology/2007/sep/26/games.news (downloaded 22 July 2010).

Johnson, Linda A. (2004). 'Video Games Used to Relax Kids in Hospital'. *Associated Press*, 9 December, http://apnews.excite.com/article/20041209/D86SBLQ00.html (downloaded 13 January 2005).

Journal of Engineering. (2012). 'Electronic Arts Inc.: Women's Olympic Ice Hockey Stars and New NHL Legends Join Roster for EA SPORTS NHL 13'. *Journal of Engineering*, p. 233.

Kaburakis, Anastasios, David A. Pierce, Beth A. Cianfrone and Amanda L. Paule. (2012). 'Is It Still "In the Game"; or Has Amateurism Left the Building? NCAA Student-Athletes' Perceptions of Commercial Activity and Sports Video Games'. *Journal of Sport Management* vol. 26 no. 4, pp. 295–308.

Kafner, Katherine M. (1986). 'Father Knows Best – Just Ask the Tramiel Boys'. *Business Week*, 15 December, pp. 106, 108.

Kane, Yukari Iwatani. (2008). 'Leading the News: Sony Again Delays PS3 Virtual Community: Service Would Help Close the Online Gap with Microsoft Xbox'.

Wall Street Journal Asia, p. 3. http://proquest.umi.com/pqdweb?did= 1869741711&Fmt=7&clientId=5258&RQT=309&VName=PQD (accessed 12 August 2013).

Kane, Yukari Iwatani and Nick Wingfield. (2010). 'Videogame Makers Look to Revive Industry'. *Wall Street Journal Asia*, p. 17. http://proquest.umi.com/pqdweb?did= 1481309221&Fmt=7&clientId=5258&RQT=309&VName=PQD (accessed 12 August 2013).

Kang, Stephanie. (2005). 'Technology – A Special Report: Bright Spot in Toyland'. *Wall Street Journal*, 23 May, p. R.13.

Kelly, Kevin. (2006). 'Gamers in Trouble: G4 Moves to E!'. *Joystiq.com*, http://www.joystiq.com/2006/10/13/gamers-in-trouble-g4-moves-to-e/ (accessed 12 August 2013).

Kent, Steven L. (2001). *The Ultimate History of Video Games: From Pong to Pokemon and Beyond – The Story behind the Craze That Touched Our Lives and Changed the World*. New York: Prima Publishing.

Kent, Steven L. (2003). 'Video Game Report Card'. *Gamespy.com*, http://www. gamespy.com/articles/491/491378p1.html (downloaded 23 February 2004).

Kerr, Aphra. (2006a). *The Business and Culture of Digital Games: Gamework/Gameplay*. London: Sage.

Kerr, Aphra. (2006b). 'Digital Games as Cultural Industry'. In *The Business and Culture of Digital Games: Gamework/Gameplay*. London: Sage, pp. 43–74.

Kerr, Aphra. (2012). 'The UK and Irish Game Industries'. In Peter Zackariasson and Timothy L. Wilson (eds). *The Video Game Industry: Formation, Present State, and Future*. New York: Routledge, pp. 116–33.

Kerr, Aphra. (2013). 'Space Wars: The Politics of Game Production in Europe'. In Nina B. Huntemann and Ben Aslinger (eds), *Gaming Globally: Production, Play and Place*. New York: Palgrave Macmillan, pp. 215–31.

Kerr, Aphra, and Roddy Flynn. (2002). *Revisiting Globalization through the Movie and Digital Game Industries*. Paper presented at the Media in Transition 2, 10–12 May, Cambridge, MA.

Kim, Ryan. (2010). 'Apple Tablet May Boost Gaming'. *San Francisco Chronicle*, p. D.1. http://proquest.umi.com/pqdweb?did=1947884611&Fmt=7&clientId= 5258&RQT=309&VName=PQD (accessed 15 July 2013).

Kim, Yun-Hee. (2011). 'Microsoft to Add Staff in China'. *Wall Street Journal (Online)*, http://ezp.bentley.edu/login?url=http://search.proquest.com/?url= http://search.proquest.com/docview/853066752?accountid=8576 (accessed 15 July 2013).

Kiss, Jemima. (2008). 'Sony's Blu-ray Wins HD DVD Battle', 19 February, http://www.guardian.co.uk/media/2008/feb/19/digitalmedia.sony?gusrc= rss&feed=technology (downloaded 3 March 2008).

Klang, Mathias. (2004). 'Avatar: From Deity to Corporate Property'. *Information, Communication and Society* vol. 7 no. 3, pp. 389–402.

Kline, Stephen, Nick Dyer-Witheford and Greg De Peuter. (2003). *Digital Play: The Interaction of Technology, Culture, and Marketing*. Montreal: McGill-Queen's University Press.

Knight-Ridder. (2006). 'IBM Ships Microchips for Nintendo's Wii Video Game Console'. *Knight Ridder Tribune Business News*, p. 1.

Konrad, Rachel. (2005). 'Programming Jobs Losing Luster in U.S.'. *Associated Press*, 21 June, http://www.highbeam.com/doc/1P1-110205972.html (accessed 13 August 2013).

Kuehr, Ruediger, German T. Velasquez and Eric Williams. (2003). 'Computers and the Environment: An Introduction to Understanding and Managing Their Impacts'. In Ruediger Kuehr and Eric Williams (eds). *Computers and the Environment: Understanding and Managing Their Impacts*. Boston, MA: Kluwer Academic Publishers, pp. 1–15

Lacina, Linda. (2012). 'Grand Theft Auto Maker Wants to Carjack the Market: The Boss of an Iconic Video Game Firm Admits He's Not a Gamer', *SmartMoney.com*, http://ezp.bentley.edu/login?url=http://search.proquest.com/docview/1011472356?accountid=8576 (accessed 15 July 2013).

LaDou, Joseph. (2006). 'Occupational Health in the Semiconductor Industry'. In Ted Smith, David A. Sonnenfeld and David Naguib Pellow (eds). *Challenging the Chip: Labor Rights and Environmental Justice in the Global Electronics Industry*. Philadelphia, PA: Temple University Press, pp. 31–42.

Lasker, John. (2008). 'Inside Africa's PlayStation War', 8 July, http://www.towardfreedom.com/home/content/view/1352/1 (downloaded 18 August 2010).

Laskowski, Nicole. (2005). 'Elementaries Offer Playstations for Supplemental Education'. *Gaylord Herald Times*, 2 March, http://articles.petoskeynews.com/2005-03-02/supplemental-education_24065937 (accessed 13 August 2013).

Learmonth, Michael and Abbey Klaassen. (2009). 'App Revenue Is Poised to Surpass Facebook Revenue'. *Advertising Age* vol. 80 no. 18, pp. 3–40.

Lee, Jenny. (2012). 'A Whole New Video Game: The Loss of Large Studios in Vancouver Has Spawned an Independent Revival'. *Vancouver Sun*, http://ezp.bentley.edu/login?url=http://search.proquest.com/docview/1037492312?accountid=8576 (accessed 15 July 2013).

Leisure & Travel Business. (2010). 'Research and Markets; Games Software in Western Europe: An Essential Resource for Top-level Data and Analysis'. *Leisure & Travel Business*, p. 66.

Leisure & Travel Week. (2009). 'Sony Computer Entertainment Inc.: New PlayStation3 Retail Sales Achieve 1 Million Units in 3 Weeks Worldwide'. *Leisure & Travel Week*, p. 40.

Levine, Robert. (2005a). 'MTV Finds a New Ally in Games'. *New York Times*, 27 June, p. C.2.

Levine, Robert. (2005b). 'A Spartan Warrior with A Sense of Humor'. *New York Times*, 21 March, p. C.7.

Levine, Robert. (2005c). 'Story Line Is Changing for Game Makers and Their Movie Deals'. *New York Times*, 21 February, p. C.1.

License! (2005). 'Comic Channels'. *License!* vol. 8 no. 4, p. 11.

Lien, Tracey (2013). 'How Middle Eastern Studios Are Overcoming Game Development Challenges'. *Polygon.com*, http://www.polygon.com/2013/7/2/4485980/how-middle-eastern-studios-are-overcoming-game-development-challenges (accessed 4 July 2013).

Loftus, Tom. (2003). '… and a Jedi Knight in "Star Wars Galaxies": The Year in Gaming'. *MSNBC.com*, 22 December, http://www.nbcnews.com/id/3703850/ns/technology_and_science-games/t/jedi-knight-star-wars-galaxies/#.Ugvnd7zpZcs (accessed 13 August 2013).

Lou, Kesten. (1999). 'PlayStation Owners Had a Better Year: Nintendo Seemed to Abandon the 64 for Pokemon Hype'. *Edmonton Journal*, p. E.9. http://proquest.umi.com/pqdweb?did=212595901&Fmt=7&clientId=5258&RQT=309&VName=PQD (accessed 15 July 2013).

Lowenstein, Douglas. (2005). 'E3 2004 State of the Union Address'. Entertainment Software Association, http://www.theesa.com/newsroom/soi_2005.asp (accessed 13 August 2013).

LucasArts. (2002). 'LucasArts Chooses Xbox and PC as Platforms for Star Wars: Knights of the Old Republic'. *LucasArts.com*, http://www.lucasarts.com/company/release/news20020228.html (accessed 13 August 2013).

Lucien, King. (2002). *Game on: The History and Culture of Video Games*. New York: Universe Publications.

Luk, Lorraine. (2011). 'Wintek Compensates Poisoned Workers'. *Wall Street Journal (Online)*, http://search.proquest.com/docview/853366164?accountid=8576 (accessed 15 July 2013).

Lüthue, Boy. (2006). 'The Changing Map of Global Electronics: Networks of Mass Production in the New Economy'. In Ted Smith, David A. Sonnenfeld and David Naguib Pellow (eds), *Challenging the Chip: Labor Rights and Environmental Justice in the Global Electronics Industry*. Philadelphia, PA: Temple University Press, pp. 17–30.

Machlup, Fritz. (1962). *The Production and Distribution of Knowledge in the United States*. Princeton, NJ: Princeton University Press.

Macmillan, Douglas. (2010). 'Facebook Gamers Are Frequent Players'. *BusinessWeek Online*, p. 14–14.

Mallory, Jordan. (2012). 'EQ 2012 Financials: Q4 Revenues and Income up Year-over-Year, 2012 Revenue up, Income down'. *Joystiq.com*, 7 May, http://www.joystiq.

com/2012/05/07/ea-2012-financials-q4-revenue-and-income-up-year-over-year-201/ (downloaded March 2013).

Market Research.com. (2004). 'Random House Wearing Halo in Game Guide Market'. *Book Publishing Report* vol. 29 no. 45, p. 6.

Market Research.com. (2006). 'eGuides Help Prima Games Thrive in Video Game Market'. *Book Publishing Report* vol. 31 no. 42, p. 5.

Market Share Reporter (2005a) 'Best Selling Video Games'. *Market Share Reporter*. Detroit, MI.

Market Share Reporter. (2005b). 'Leading Graphics Chips Makers Worldwide, 2005'. *Market Share Reporter*. Detroit, MI.

Market Share Reporter. (2005c). 'Strategy Guide Industry'. *Market Share Reporter*. Detroit, MI.

Market Share Reporter. (2007). 'Leading Desktop Graphics Chips Makers, 2006'. *Market Share Reporter.* Detroit, MI.

Market Share Reporter. (2008a). 'Top Software Firms Worldwide, 2007'. *Market Share Reporter*. Detroit, MI.

Market Share Reporter. (2008b). 'Desktop Graphic Market Worldwide, 2007'. *Market Share Reporter*. Detroit, MI.

Market Share Reporter. (2009a). 'Leading Desktop GPU Market Worldwide, 2008'. *Market Share Reporter*. Detroit, MI.

Market Share Reporter. (2009b). 'Top Video Game Publishers, 2008'. *Market Share Reporter*. Detroit, MI.

Market Share Reporter. (2009c). 'Top Video Game Retailers, 2008'. *Market Share Reporter*. Detroit, MI.

Market Share Reporter. (2009d). 'Used Video Game Market, 2007'. *Market Share Reporter*. Detroit, MI.

Market Share Reporter. (2010a). 'Top Graphics Chips Makers Worldwide, 2009'. *Market Share Reporter*. Detroit, MI.

Market Share Reporter. (2010b). 'Video Game Console Market in Russia, 2009'. *Market Share Reporter*. Detroit, MI.

Market Share Reporter. (2010c). 'Video Game Industry in Australia, 2009'. *Market Share Reporter*. Detroit, MI.

Market Share Reporter. (2010d). 'Video Game Industry in Canada, 2009'. *Market Share Reporter*. Detroit, MI.

Market Share Reporter. (2010e). 'Video Game Industry in France, 2009'. *Market Share Reporter*. Detroit, MI.

Market Share Reporter. (2010f). 'Video Game Industry in India, 2009'. *Market Share Reporter*. Detroit, MI.

Market Share Reporter. (2010g). 'Video Game Industry in Italy, 2009'. *Market Share Reporter*. Detroit, MI.

Market Share Reporter. (2010h). 'Video Game Industry in Japan, 2009'. *Market Share Reporter*. Detroit, MI.

Market Share Reporter. (2010i). 'Video Game Industry in the Middle East, 2009'. *Market Share Reporter*. Detroit, MI.

Marketingvox.com (2006) 'Nielsen: 56% of Active Gamers Are Online, 64% Are Women', 5 October, http://www.marketingvox.com/nielsen_56_of_active_gamers_are_online_64_are_women-022774/ (downloaded 11 May 2007).

Markoff, John. (2003). 'From PlayStation to Supercomputer for $50,000'. *New York Times*, 26 May, p. C.3.

Markoff, John. (2006). 'IBM in Video-chip Deal Products to Be Used in Portable Devices'. *International Herald Tribune*, p. 15, http://proquest.umi.com/pqdweb?did=984947841&Fmt=7&clientId=5258&RQT=309&VName=PQD (accessed 15 July 2013).

Marr, Merissa. (2005). 'Videogames Grow up: Disney Gets into the Game with Two New Industry Deals; Eyeing Both Adults and Kids'. *Wall Street Journal*, 19 April, p. B.1.

Marriott, Michel. (2005a). 'In Console Wars, Xbox Is Latest to Rearm'. *New York Times*, 13 May.

Marriott, Michel. (2005b). 'On Screens, but Not Store Shelves: Casual Games'. *New York Times*, 27 June, http://www.nytimes.com/2005/06/27/technology/27casual.html?_r=0 (accessed 13 August 2013).

Martin, Matt. (2009). 'THQ Has 50/50 Chance of Going Bankrupt'. *Gameindustry.biz*, http://www.gamesindustry.biz/articles/thq-has-50-50-chance-of-going-bankrupt-hickey (accessed 12 August 2013).

Martin, Scott. (2008). 'GameStop Gobbles French Video Game Chain'. *Red Herring*, p. 4.

Martin, Stana. (2002). 'The Political Economy of Women's Employment in the Information Sector'. In Eileen Meehan and Ellen Riordan (eds). *Sex & Money: Feminism and Political Economy in the Media*. Minneapolis: University of Minnesota Press, pp. 75–87.

Matthews, Matt. (2009a). 'In Depth: The State of GameStop, Part One'. *Gamasutra.com*, 28 April, http://www.gamasutra.com/news?story=23357 (downloaded 1 March 2010).

Matthews, Matt. (2009b). 'In Depth: GameStop Controls 21 Percent of U.S. Game Market'. *Gamasutra.com*, 1 May, http://www.gamasutra.com/php-bin/news_index.php?story=23425 (downloaded 1 March 2010).

Maxwell, Richard and Toby Miller. (2012). *Greening the Media*. New York: Oxford University Press.

Mazurek, Jan. (2003). *Making Microchips: Policy, Globalization, and Economic Restructuring in the Semiconductor Industry*. Cambridge, MA: MIT Press.

McCallister, Ken. (2005). *Game Work: Language, Power, and Computer Game Culture*. Tuscaloosa: University of Alabama Press.

McCammon, Holly J. and Larry J. Griffin. (2000). 'Workers and Their Customers and Clients'. *Work and Occupations* vol. 27 no. 3, August, pp. 278–93.

McCune, Jenny C. (1998). 'The Game of Business'. *American Management Journal*, February, pp. 56–8.

McGonigal, Jane. (2011). *Reality Is Broken: Why Games Make Us Better and How They Can Change the World*. New York: Penguin.

McNary, Dave. (2004). 'Par Finds Clancy Vidgame Easy 'Cell''. *Variety*, 16 December, p. 1.

McQuade, Walter. (1979). 'In Games and Gambling, a Good Time Was Had by Almost All'. *Fortune* vol. 99, 18 June, p. 186.

McWhertor, Michael. (2009). 'Report: Game Industry Spent $823M Shilling Games in '09'. *Kotaku.com*, http://kotaku.com/5150987/report-game-industry-spent-823m-shilling-games-in-08 (downloaded 13 June 2010).

Media. (2009). 'Wii Needs to Prove It Can Be More than Just a Fad'. *Media*, p. 14.

Meehan, Eileen. (2002). 'Gendering the Commodity Audience: Critical Media Research, Feminism, and Political Economy'. In Eileen R. Meehan and Ellen Riordan (eds). *Sex & Money: Feminism and Political Economy in the Media*. Minneapolis: University of Minnesota Press, pp. 209–22.

Mergent. (2008). 'Corporate Profiles', http://www.mergent.com (downloaded 22 July 2008).

Microsoft. (2005). *Annual Report, 2004*. Redmond, WA: Microsoft, Inc.

Microsoft. (2007). *Annual Report, 2006*. Redmond, WA: Microsoft, Inc.

Miege, Bernard. (1989). *The Capitalization of Cultural Production*. New York: International General.

Miller, Patrick. (2012). 'Eleventh Annual Salary Survey'. *Game Developer* vol. 19, pp. 7–13.

Miller, Toby. (2008). 'Anyone for Games? Via the New International Division of Cultural Labor'. In Helmut Anheier and Yudhishthir Raj Isar (eds). *The Cultural Economy*. London: Sage, vol. 2, pp. 227–40.

Miller, Trudy. (2005). 'Telephone Interview'. Redwood City, California.

Moffat, Susan. (1990). 'Can Nintendo Keep Winning?'. *Fortune* vol. 122, pp. 131–2, 136.

Moledina, Jamil. (2004a). 'The Art of the Game Deal – Expanded Edition'. *Gamsutra.com*, 2005, 8 October, http://www.gamasutra.com/view/feature/130559/the_art_of_the_game_deal_expanded_.php?print=1 (accessed 13 August 2013).

Moledina, Jamil. (2004b). 'Letter from the Editor'. *Game Developer's Fall 2004 Game Career Guide*, p. 66.

Moltenberry, Karen. (2006). 'Casual Approach'. *Computer Graphics World* vol. 29, April, p. 44.

Morris, Chris. (2003). 'Constitution Protects Video Games'. *CNNMoney*, http://money.cnn.com/2003/06/03/technology/games_firstamendment/ (downloaded 11 May 2005).

Moscaritolo, Angela. (2012). 'Call of Duty Online Heading to China'. *PC Magazine*, p. 1.

Mosco, Vincent. (1999). *Lost in Space*. Paper presented at the Union for Democratic Communications, Eugene, Oregon, 16 October.

Moss, Richard. (2013). 'Big Game: The Birth of Kenya's Games Industry'. *Polygon.com*, http://www.polygon.com/features/2013/7/3/4483276/kenya-games-industry (accessed 4 July 2013).

MPAA. (2009). *Theatrical Market Statistics, 2008*. Motion Picture Association of America, http://www.stop-runaway-production.com/wp-content/uploads/2009/07/2008-MPAA-Theatrical-Stats.pdf (accessed 12 August 2013).

MPAA. (2012). *Theatrical Market Statistics, 2011*. Motion Picture Association of America, http://www.mpaa.org/resources/5bec4ac9-a95e-443b-987b-bff6fb5455a9.pdf (accessed 12 August 2013).

Musgrove, Mike. (2001). 'PlayStation 2 Has Competitors Playing Catch-up'. *Washington Post*, p. E.01. http://proquest.umi.com/pqdweb?did=96644694&Fmt=7&clientId=5258&RQT=309&VName=PQD (accessed 15 July 2013).

National Post. (2009). 'Game Is Afoot at Apple; iPod a Golden Opportunity for Video-game Developers'. *National Post*, p. FP.3, http://proquest.umi.com/pqdweb?did=1865880261&Fmt=7&clientId=5258&RQT=309&VName=PQD (accessed 15 July 2013).

Neff, Robert and Maria Shao. (1990). 'The Newest Nintendo Will Take a Slow Boat to America'. *Business Week*, 2 July, p. 46.

Nelson, Robin. (1990a). 'A Video Game That Tracks Stocks, Too'. *Popular Science* vol. 237, December, p. 93.

Nelson, Robin. (1990b). 'Video Games Aim at Reality'. *Popular Science* vol. 237, December, pp. 90–3.

Netsel, Tom. (1990). 'The Design Game'. *Compute* vol. 11, pp. 76–8, 80.

New York Times. (2002). 'Technology Briefing: Hardware'. *New York Times*. 2 July, p. 11

New York Times, (2005a). 'GameStop to Acquire Rival Video Game Retailer'. *New York Times*, 19 April, p. C.9.

New York Times. (2005b). 'N.B.A. to Announce Deals with Five Video Game Publishers'. *New York Times*, 22 March, http://www.nytimes.com/2005/03/22/technology/22game.html (accessed 13 August 2013).

New Zealand Herald. (2003). 'Computer Game Banned for Repetitive Extreme Violence'. *New Zealand Herald*, 12 December, http://www.nzherald.co.nz/nz/news/article.cfm?c_id=1&objectid=3539142 (accessed 13 August 2013).

Newman, Heather. (2005). 'GAME BITS: '04 Sales Prove That PS2 and Xbox Are Aging Well'. *Knight-Ridder Tribune Business News*, 23 January, p. K3816.

Newman, James. (2002). 'In Search of the Videogame Player: The Lives of Mario'. *New Media & Society* vol. 4 no. 3, pp. 405–22.

Newman, Jared. (2011). 'Valve Plots Living Room Takeover with Steam', 1 March, http://newsfeed.orgfree.com/article.html?http://feedproxy.google.com/~r/Playstation3/~3/qunuT2tJWQ4/url (downloaded 3 March 2011).

Newsweek. (1989). 'A Game of Legal Punch-out'. *Newsweek* vol. 113, 2 January p. 50.

Nichols, Mark. (1988). 'Atari's Odd Man Out'. *Maclean's* vol. 101, 25 January: 25.

Nichols, Randy. (2005). *The Games People Play: A Political Economic Analysis of Video Games and Their Production*. Unpublished dissertation, University of Oregon, Eugene.

Nichols, Randy. (2008). 'Ancillary Markets: Merchandising and Video Games'. In Janet Wasko and Paul McDonald (eds). *The Contemporary Hollywood Film Industry*. Malden, MA: Blackwell Publishers.

Nichols, Randy. (2009). 'Target Acquired: America's Army and the Political Economy of the Global Video Games Industry'. In Nina B. Huntemann and Matthew Thomas Payne (eds). *Joystick Soldiers: The Politics of Play in Military Video Games*. New York: Routledge.

Nichols, Randy. (2013). 'Who Plays, Who Pays? Mapping Video Game Production and Consumption Globally'. In Nina B. Huntemann and Ben Aslinger (eds). *Gaming Globally: Production, Play and Place*. New York: Palgrave Macmillan, pp. 19–39

Nieborg, David B. (2009). 'Training Recruits and Conditioning Youth: The Soft Power of Military Games'. In Nina B. Huntemann and Matthew Thomas Payne (eds). *Joystick Soldiers: The Politics of Play in Military Video Games*. New York: Routledge, pp. 53–66.

Nielsen. (2005). *Video Gamers in Europe* (Report). Brussels: Interactive Software Federation of Europe, http://www.lateledipenelope.it/public/ISFE05_ConsumerStudy.pdf (downloaded 12 August 2013).

Nielsen. (2008a). 'US TV Households up 1.5% – Asian, Hispanic Households Triple That', 8 September, http://www.marketingcharts.com/television/us-tv-households-up-15-asian-hispanic-households-triple-that-5846/ (downloaded 15 September 2010).

Nielsen. (2008b). *Video Gamers in Europe* (Report). Brussels: Interactive Software Federation of Europe, http://www.isfe-eu.org/tzr/scripts/downloader2.php?filename=T003/F0013/8c/79/w7ol0v3qaghqd4ale6vlpnent&mime=application/pdf&originalname=ISFE_Consumer_Research_2008_Report_final.pdf (accessed 12 August 2013).

Niizumi, Hirohiko. (2004). 'Sales down, Number of Gamers up in Japan', 26 July, http://www.gamespot.com/news/6103414.html (downloaded 8 July 2005).

Nintendo. (2005a). *Annual Report, 2004*. Kyoto: Nintendo, Ltd.

Nintendo (2005b). *Company History*. Kyoto: Nintendo Ltd, http://www.nintendo. com/corp/history.jsp (downloaded 20 August 2005).

Nintendo. (2008a). *Annual Report, 2007*. Kyoto: Nintendo, Ltd.

Nintendo (2008b). *Company History*. Kyoto: Nintendo Ltd http://www.nintendo. com/corp/history.jsp (downloaded 22 July 2008).

Nolan, Rachel. (2010). 'Virtual World Order'. *Boston Globe*, p. K.4. http://ezp.bentley. edu/login?url=http://search.proquest.com/docview/818574604?accountid=8576 (accessed 12 August 2013).

NPD. (2002). 'Annual 2001 Video Game Best Selling Titles', http://www.npdfunworld. com/fun/Servlet?nextpage=trend_body.html&content_id=287 (downloaded 27 September 2005).

NPD. (2003). 'Fourth Quarter 2002 Video Games Best Sellers', http://www. npdfunworld.com/fun/Servlet?nextpage=trend_body.html&content_id=352 (downloaded 27 September 2005).

NPD. (2004). 'Top 10 Video Game Titles Ranked by Total U.S. Units Annual 2003', http://www.npdfunworld.com/fun/Servlet?nextpage=trend_body.html&content_ id=780 (downloaded 27 September 2005).

NPD. (2005). 'Top 10 Video Game Titles Ranked by Total U.S. Units Annual 2004', http://www.npdfunworld.com/fun/Servlet?nextpage=trend_body.html&content_ id=2079 (downloaded 27 September 2005).

Nuttall, Chris. (2005). 'NBA Signs Non-exclusive Deals Video Games'. *Financial Times*, p. 30.

Obermayer, Joel. (2000). 'Blue-collar Tech'. *Industry Standard*, 5 June, http://www.thestandard.com/article/display/0,1153,15606,00.html (accessed 17 June 2000).

O'Connor, Susan. (2003). 'How to Satisfy Women', http://www.igda.org/articles/ socconnor_women.php (downloaded 10 June 2005).

OFLC. (2008). 'Film Label: Censorship', http://www.censorship.govt.nz/censorship-film-labels.html – Classification (downloaded 8 July 2008).

Olenick, Doug. (1999). 'Dreamcast Has Registers Ringing'. *TWICE* vol. 14 no. 29, p. 68.

Olsen, Jennifer. (2001). *2001 Game Developer Salary Survey*. *Gamasutra.com*, 15 July, http://www.gamasutra.com/view/feature/131465/game_development_salary_survey_ 2001.php (accessed 12 August 2013).

Olsen, Jennifer. (2002). *2002 Game Developer Salary Survey*. *Gamasutra.com*, 15 July, http://www.gamasutra.com/view/feature/131392/game_development_salary_survey_ 2002.php (accessed 12 August 2013).

Olsen, Jennifer. (2003). *2003 Game Developer Salary Survey*. *Gamasutra.com*, 10 February, http://www.gamasutra.com/view/feature/130444/game_development_ salary_survey_2003.php (accessed 12 August 2013).

Orland, Kyle. (2010). 'Greenpeace Criticizes Nintendo and Microsoft's Environmental Records', 26 October, http://www.gamasutra.com/view/news/ 31174/Greenpeace_Criticizes_Nintendo_and_Microsofts_Environmental_Records. php (downloaded 11 February 2011).

Oser, Kris. (2004). 'Microsoft's Halo2 Soars on Viral Push'. *Advertising Age* vol. 75 no. 43, 25 October, http://adage.com/article/news/microsoft-s-halo2-soars-viral-push/100896/ (accessed 13 August 2013).

Patel, Kunur. (2010). 'All the World's a Game, and Brands Want to Play Along'. *Advertising Age* vol. 81 no. 22, p. 4.

Pavlik, John V. (1998). *New Media Technology: Cultural and Commercial Perspectives*, 2nd edn. Boston, MA: Allyn & Bacon.

Peckham, Matt. (2008). '2007 Video Game Sales Soar by Record-Shattering 43%'. *PCWorld.com*, 17 January, http://blogs.pcworld.com/gameon/archives/006324.html (downloaded 1 July 2008).

Peddle, John. (2014). E-mail Interview. San Francisco, CA.

PEGI. (2007). 'Pan European Game Information – Ratings Explained', http://www. pegi.info/en/index/id/176 (downloaded 8 July 2008).

People. (1985). 'How a Computer-game Maker Finesses the Software Slow-down'. *People*, 10 June, p. 72.

Pereira, Joseph. (2002a). 'A Season of Peace, Love – and Decapitation'. *Wall Street Journal*, 17 December, p. B.1.

Pereira, Joseph. (2002b). 'Showdown in Mario Land – Game Makers Create Deluge of Titles as Microsoft Makes It a Three-Way Race'. *Wall Street Journal*, 19 April, p. A.13.

Peréz Fernández, Agustín (2013). 'Snapshot 2: Video Game Development in Argentina'. In Nina B. Huntemann and Ben Aslinger (eds). *Gaming Globally: Production, Play and Place*. New York: Palgrave Macmillan, pp. 79–81.

Perez Martin, D. Joaquin, Julio Ignacio Ruiz and Sonia Portillo Martinez. (2006). *Women and Video Games: Habits and Preferences of the Video Gamers*. Madrid: Universidad Europea de Madrid, http://www.isfe-eu.org/tzr/scripts/downloader2. php?filename=T003/F0013/6d/e1/bbbf9e934586e4fc34bde678281f57f0&mime= application/pdf&originalname=womenandgames.pdf (accessed 27 May 2007).

Pescovitz, David. (1997). 'The Cutting Edge/Cybernews: Sega Is Seeking Just a Little Console Elation; Games: Despite Being Trounced by Competitors, the Video Toy Maker Says It's Not Pulling the Plug Yet'. *Los Angeles Times*, p. 3, http://proquest. umi.com/pqdweb?did=12454362&Fmt=7&clientId=5258&RQT=309&VName= PQD (accessed 15 July 2013).

Peterson, Thane and Maria Shao. (1990). 'But I Don't Wanna Play Nintendo Anymore!' *Business Week*, 19 November, pp. 52, 54.

Pfleger, Katherine. (2001). 'Unions Address High-tech Workers', 28 January, http://news. excite.com/news/ap/010128/13/high-tech-unions (downloaded 28 January 2001).

Pitta, Julie. (1990). 'This Dog Is Having a Big Day'. *Forbes* vol. 145, 22 January, pp. 106–7.

Plunkett, Luke. (2010). 'What's All This "PlayStation Wars" Business?'. *Kotaku.com*, 25 June, http://kotaku.com/5028998/whats-all-this-playstation-wars-business (accessed 15 July 2011).

Porat, Marc Uri. (1977). *The Information Economy: Definition and Measurement* (OT special publications, 77–12). Washington, DC: US Department of Commerce, Office of Telecommunications.

Portnow, James, Arthur Protasio and Kate Donaldson. (2013). 'Snapshot 1: Brazil: Tomorrow's Market'. In Nina B. Huntemann and Ben Aslinger (eds). *Gaming Globally: Production, Play and Place*. New York: Palgrave Macmillan, pp. 75–7.

Postigo, Hector. (2003). 'From Pong to Planet Quake: Post-industrial Transitions from Leisure to Work'. *Information, Communication and Society* vol. 6 no. 4, pp. 593–607.

Pöyhönen, Päivi and Eeva Simola. (2007). *Connecting Components, Dividing Communities*. Helsinki: FinnWatch.

PR Newswire. (2009). 'Gameloft: 20% Growth in Sales over the First Half of 2009'. *PR Newswire*.

PR Newswire. (2010a). 'Mobile Gaming Lags among Consumers on Android Platform'. *PR Newswire*.

PR Newswire. (2010b). 'PlayStation(R) Network Fuels the Next Evolution of Digital Distribution'. *PR Newswire*.

Pratchett, Rhianna. (2005). *Games in the UK: Digital Play, Digital Lifestyles*. London: BBC, http://open.bbc.co.uk/newmediaresearch/files/BBC_UK_Games_Research_2005.pdf (accessed 27 May 2006).

Prima Games. (2012). 'About Us', http://www.primagames.com/pages/about-us (downloaded 4 October 2012).

Pringle, David. (2002). 'Cellphone Giant Nokia Plans Videogame Console'. *Wall Street Journal*, 5 November, p. D4.

Provenzo, Jr, Eugene F. (1991). *Video Kids: Making Sense of Nintendo*. Cambridge, MA: Harvard University Press.

Race, Tim. (2001). 'New Economy: Unionization Drives at Dot-coms Have a Familiar Ring'. *New York Times.com*, 22 January, http://www.nytimes.com/2001/01/22/technology/22NECO.html (downloaded 25 January 2001).

Rahim, Saquib. (2010). 'Video Gamers Use as Much Energy as San Diego'. *Scientific American*, 17 December, http://www.scientificamerican.com/article.cfm?id=video-gamers-use-as-much-energy-as-san-diego (accessed 12 August 2013).

Reed, Sandra R. and Cheryl Spencer. (1986). 'James H. Levy on Reaching More People with Computing'. *Personal Computing* vol. 10, pp. 145, 147, 149, 151.

Reuters. (2000a). '28 States Sue Record Labels Claiming Price Fixing'. *CNN.com*,
    8 August, http://www.cnn.com/2000/LAW/08/08/media.compactdiscs.lawsuit.reut/
    index.html (downloaded 8 August 2001).

Reuters. (2000b). 'Universal Music Confirms Digital Download Trials'. *New York
    Times.com*, 1 August, http://www.nytimes.com/reuters/technology/tech-media-
    universal.html (downloaded 3 August 2001).

Reuters. (2005a). 'Indian Software Testing Moves from Boredom to Boo. *New York
    Times.com*, 31 March. (downloaded 1 April 2005).

Reuters. (2005b). 'Sony Says to Narrow Focus of R&D'. *New York Times.com*, 23 June
    (accessed 24 June 2005).

RIAA. (2000). 'The World Sound Recording Market', http://www.riaa.org/MD-
    World.ctm (downloaded 29 August 2000).

Rice, Berkeley. (1979). 'How Electronic Games Grew up'. *Psychology Today* vol. 13,
    November, pp. 97–8.

Richtel, Matt. (2004). 'Game Sales Thrive Thanks to the Big Kids (in Their 20's)'.
    New York Times, 27 December, p. C.1.

Richtel, Matt. (2005a). 'At This Restaurant, the Video Games Come with the Meal'.
    New York Times, 30 May, http://www.nytimes.com/2005/05/30/technology/
    30bushnell.html (accessed 13 August 2013).

Richtel, Matt. (2005b). 'Fringes vs. Basics in Silicon Valley'. *New York Times*, 9 March,
    p. C.1.

Richtel, Matt. (2005c). 'Just Past Midnight, the Game Is in Hand'. *New York
    Times.com*, 25 March, http://www.nytimes.com/2005/03/25/business/
    25sony.html?_r=0 (downloaded 25 March 2005).

Richtel, Matt. (2005d). 'A New Reality in Video Games: Advertisements'. *New York
    Times*, p. C.1.

Richtel, Matt. (2005e). 'Unions Struggle as Communications Industry Shifts'. *New
    York Times*, 1 June, p. C.1.

Robischon, Noah. (2005). 'Introducing 'Dirty Harry,' via Video Game, to a New
    Generation'. *New York Times*, 28 February, p. C.6.

Roch, Stuart. (2004). 'The New Studio Model'. *Gamasutra.com*, 29 October.
    http://www.gamasutra.com/view/feature/130568/the_new_studio_model.php
    (downloaded 25 March 2005).

Roe, Keith and Daniel Muijs. (1998). 'Children and Computer Games: A Profile of
    the Heavy User'. *European Journal of Communication* vol. 13 no. 2, pp. 354–72.

Rogers, Michael. (1990). 'Nintendo and Beyond'. *Newsweek* vol. 115, p. 63–63.

Rooney, Megan. (2003). 'The Games People Don't Play'. *Chronicle of Higher
    Education*, 11 July, p. A.9.

Rose, Mike. (2013). 'Asia, Europe Will Make up 87% of Global Mobile, Online Game
    Revenue in 2015 – Report', 15 January, http://www.gamasutra.com/view/news/

184834/Asia_Europe_will_make_up_87_of_global_mobile_online_game_revenue_ in_2015__report.php (downloaded 23 March 2013).

Rosenberg, Dave. (2010). 'Korea Rules Virtual Currency as Good as Cash'. *CNET.com*, 19 January, http://news.cnet.com/8301-13846_3-10437250-62.html (downloaded 10 October 2012).

Rosenbloom, Stephanie. (2005). 'New York and Boston', *New York Times.com*, 12 May, p. 328 (downloaded 13 May 2005).

Ross, Andrew. (1999). 'Sweated Labor in Cyberspace'. *New Labor Forum*, Spring/Summer, pp. 47–56.

Ruggill, Judd. (2004). 'Corporate Cunning and Calculating Congressmen: A Political Economy of the Game Film'. *TEXT Technology* vol. 13 no. 1, pp. 53–72.

Rusak, Gary. (2009). 'iPhone therefore iApp'. *KidScreen*, p. 15.

Salsberg, Art. (1977). 'TV Electronic Games Grow up'. *Popular Electronics* vol. 12, September, p. 4.

Sandqvist, Ulf. (2012). 'The Development of the Swedish Game Industry: A True Success Story?' In Peter Zackariasson and Timothy L. Wilson (eds). *The Video Game Industry: Formation, Present State and Future*. New York: Routledge, pp. 134–153.

Sarkar, Samit. (2013). 'Microsoft Posts 29 Percent Year-over-year Drop in Quarterly Xbox 360 Platform Revenue'. *Polygon.com*, 24 January, http://www.polygon. com/2013/1/24/3912722/microsoft-q2-2013-fiscal-report-xbox-360-revenue-drop (downloaded 23 March 2013).

Sarno, David. (2010). 'How I Made It: Michael Morhaime: He's Serious about Games'. *Los Angeles Times*, p. B.2, http://proquest.umi.com/pqdweb?did= 2103617071&Fmt=7&clientId=5258&RQT=309&VName=PQD (accessed 12 August 2013).

Satariano, Adam. (2010). 'Did Activision Just Frag Itself?' *Business Week*, p. 53, http://www.businessweek.com/magazine/content/10_18/b4176053952018.htm? chan=magazine+channel_news+-+technology%3E)/ (accessed 12 August 2013).

Scelsi, Chrissie. (2010). 'You Let My Avatar Do What? Recent Developments in Music Video Games'. *Entertainment and Sports Lawyer* vol. 28 no. 1, pp. 25–7.

Schiesel, Seth. (2005a). 'Retro Redux: When Yellow Blobs Ate Other Yellow Blobs'. *New York Times*, 6 April, p. E.1.

Schiesel, Seth. (2005b). 'They Got (Video) Game: N.B.A. Finals Can Wait'. *New York Times*, 21 June, p. A.1.

Schiesel, Seth. (2006). 'The Land of the Video Geek'. *New York Times*, 8 October, p. 2.1.

Schiffmann, William. (2002). 'Classes Cultivate Game Developers'. *Associated Press*, 31 March, http://www.apnewsarchive.com/2002/Classes-Cultivate-Game-Developers/id-6abf31ac3f0acf042ff301637dbec67f (accessed 13 August 2013).

Schmerken, Ivy. (2008). 'Game on – Wall Street Is Exploring Graphic Processor Units Found in Video Games to Speed Options Analytics and Other Math-intensive Applications'. *Wall Street & Technology* vol. 26 no. 7, p. 17.

Sega. (2007). *Annual Report*. Tokyo: Sega-Sammy Holdings.

Sega. (2010). 'Sega History', http://www2.sega.com/corporate/corporatehist.php (downloaded 11 May 2010).

Seitz, Patrick. (2010). 'Casual Game Firm PopCap Is Making Noise Success of "Bejeweled" and Other Games Has It Thinking IPO Is Possible.' *Investor's Business Daily*, 29 March, p. A05.

Shao, Maria. (1988). Jack Tramiel Has Atari Turned Around – Halfway. *Business Week*, 20 June, pp. 50, 52.

Shao, Maria. (1989). 'There's a Rumble in the Video Game Arcade'. *Business Week*, 20 February, p. 37.

Shapira, Ian. (2011). 'Online Gaming Boon to Asian Workers'. *Washington Post*, p. A.16.

Shaw, Adrienne. (2013). 'How Do You Say Gamer in Hindi? Exploratory Research on the Indian Digital Game Industry and Culture'. In Nina B. Huntemann and Ben Aslinger (eds). *Gaming Globally: Production, Play and Place*. New York: Palgrave Macmillan, pp. 183–201.

Shaw, Keith. (2010). 'Xbox 360 vs. PS3 vs. Wii.' *Network World (Online)*, 7 June, http://www.networkworld.com/news/2010/060710-xbox-360-vs-ps3-vs.html (accessed 13 August 2013).

Shayndi, Raice. (2012). Zynga CEO Issues Options to All Employees. *Wall Street Journal (Online)*, 9 August, http://online.wsj.com/article/SB10000872396390 44340400457757975252511042894.html (accessed 13 August 2013).

Sheff, David. (1993). *Game Over: How Nintendo Zapped an American Industry, Captured Your Dollars, and Enslaved Your Children*. New York: Random House.

Sheffield, Brandon. (2009). 'Download Revolution'. *Game Developer* vol. 16 no. 9, p. 21.

Sheffield, Brandon and Jeffrey Fleming. (2010). 'Ninth Annual Salary Survey'. *Game Developer* vol. 17 no. 7, pp. 7–13.

Sherr, Ian. (2012). 'Videogame Start-up on Live Hits Reset Button'. *Wall Street Journal*, 20 August, p. B.4.

Sherwin, Adam. (2012). 'Tax Deal for Shoot-'em-ups (as Long as They're "British")'. *Independent*, 4 October, p. 19.

Shields, Mike. (2010). 'Group Dynamics'. *Adweek* vol. 51 no. 10, p. 8.

Shirinian, Ara. (2012). '10 Years of Salary Surveys'. *Game Developer* vol. 19 no. 2, p. 7.

Sieberg, Daniel. (2002). '24-hour Video Game Channel Set to Launch'. *CNN.com*, 12 April, http://archives.cnn.com/2002/TECH/ptech/04/12/video.game.channel/ (accessed 13 August 2013).

Šisler, Vit. (2013). 'Video Game Development in the Middle East: Iran, the Arab World, and Beyond'. In Nina B. Huntemann and Ben Aslinger (eds). *Gaming Globally: Production, Play and Place*. New York: Palgrave MacMillan, pp. 251–71.

Siwek, Stephen E. (2010). *Video Games in the 21st Century: The 2010 Report*, Entertainment Software Association, http://www.theesa.com/facts/pdfs/VideoGames21stCentury_2010.pdf (accessed 13 August 2013).

Slagle, Matt. (2005a). 'Handheld Video Game Battle Heats up'. *Associated Press*, 5 January, http://apnews.excite.com/article/20050105/D87E75C80.html (downloaded 17 January 2005).

Slagle, Matt. (2005b). 'Next-gen Consoles Expected to Star in Annual Video Game Confab'. *USA Today*, 15 May, http://usatoday30.usatoday.com/tech/products/games/2005-05-15-e3-preview_x.htm (accessed 13 August 2013).

Slagle, Matt. (2005c). 'Sony Set to Unveil New PlayStation 3'. *Associated Press*, 16 May, http://www.apnewsarchive.com/2005/Sony-Set-to-Unveil-New-PlayStation-3/id-f2479cc5211c3e8212f7479d95f677ee (accessed 13 August 2013).

Sloan, Scott. (2009). 'Get the Inside Story of the Development of New Video Game Chips'. *McClatchy – Tribune News Service*.

Smith, Ethan. (2000). 'Labor Pains for the Internet Economy'. *Industry Standard*, 20 October, http://www.thestandard.com/article/display/0,1153,19555,00.html (downloaded 12 December 2000).

Smith, Nicola. (2010). 'Play Group'. *New Media Age*, pp. 21–2.

Snapshot Series. (2003). 'US Games Software & Hardware 2003'. *Snapshot Series*, p. 1.

Snapshot Series. (2004). 'US Games Software & Hardware 2004'. *Snapshot Series*, p. 1.

Snider, Mike. (2004). 'Strategy Guides Are as Hot as Video Games'. *USA Today*, 10 December, http://usatoday30.usatoday.com/life/books/news/2004-12-09-gaming-books_x.htm (accessed 13 August 2013).

Snider, Mike. (2010). 'Publishers Try to Open up More Lines of Income', *USA Today*, 9 August, p. D.3.

Snider, Mike. (2012). 'Fewer Americans Playing Video Games'. *Chicago Post-Tribune*, 6 September, http://posttrib.suntimes.com/business/14973932-420/fewer-americans-playing-video-games.html (accessed 13 August 2013).

Snider, Mike and Steven Kent. (2005). 'Gaming Hits Next Level with Xbox 360'. *USA Today*, 12 May, http://usatoday30.usatoday.com/tech/products/games/2005-05-12-xbox-revealed_x.htm (downloaded 4 January 2005).

Snow, Blake. (2009). 'Sony Tries to Boost PS3 Development with Dev Kit Price Cut'. *Arstechnica.com*, 23 March, http://arstechnica.com/gaming/2009/03/sony-announces-lower-cost-ps3-dev-tools/ (downloaded 10 October 2010).

Sony. (2005). *Annual Report, 2004*. Tokyo: Sony Corporation.

Sony. (2006). *Annual Report, 2005*. Tokyo: Sony Corporation.

Sony. (2008). *Annual Report, 2007*. Tokyo: Sony Corporation.

Sousa, Catarina. (2010). *Logics of a Transborder Communication Process. Case Study: Importation Process of the Radio Program 'Janela Aberta', Broadcast on Rádio Clube Português*. Paper presented at the International Association for Media and Communication Research.

Spectrum Strategy Consultants. (2001). *From Exuberant Youth to Sustainable Maturity: Competitive Analysis of the UK Games Software Sector*. London: Spectrum Strategy Consultants, http://dti.gov.uk/cii/services/content/industry/computer_games_leisure_software.shtml (accessed 1 April 2005).

Stanley, T. L. (2004a). 'Joystick Nation'. *Advertising Age* vol. 75, 22 March, pp. 1, 29.

Stanley, T. L. (2004b). 'Marketers Flock to Gaming Gathering'. *Advertising Age* vol. 75, 17 May, pp. 4, 158.

Steiert, Robert. (2006). 'Unionizing Electronics: The Need for New Strategies'. In Ted Smith, David A. Sonnenfeld and David Naguib Pellow (eds). *Challenging the Chip: Labor Rights and Environmental Struggle in the Global Electronics Industry*. Philadelphia, PA: Temple University Press, pp. 191–200.

Stelter, Brian. (2010). 'Xbox Console Makers Focus on Other Media and Uses'. *Ledger*, 28 January, http://www.theledger.com/article/20100128/NEWS/1285051 (accessed 13 August 2013).

Stone, Mary. (2010). 'Game Could Be Big Advance for Small Screen'. *Rochester Business Journal* vol. 26 no. 9, p. 7.

Stone, Toby, Jesse Belgrave and Srdjan Kovacevic. (2006). *The Games Industry in Eastern Europe*. London: Department of Trade and Industry, http://www.ictventuregate.eu/wp-content/uploads/2012/01/file35243.pdf (accessed 13 August 2013) .

Strong, Everard. (2003). 'Step 1: Makes Games, Step 2: Don't Suck'. *Game Developer* vol. 10, July, p. 12.

Subrahmanyam, Kaveri, Patrica M. Greenfield, Justine Cassell and Henry Jenkins. (1999). 'Computer Games for Girls: What Makes Them Play'. In *From Barbie to Mortal Kombat: Gender and Computer Games*. Cambridge, MA: MIT Press, pp. 46–71.

Suellentrop, Chris. (2012). 'Big Franchise Modern Warfare'. *Rolling Stone*, 31 May, pp. 70, 72.

Sunday Times. (2012). '$20,000 Kick-start'. *Sunday Times*, p. 20, http://ezp.bentley.edu/login?url=http://search.proquest.com/docview/1113551070?accountid=8576 (accessed 13 August 2013).

Swett, Clint. (2003). 'Saying No to Games'. *Sacramento Bee*, 8 December, http://www.sacbee.com/content/business/story/7920261p-8857902c.html (accessed 19 December 2003).

Tabuchi, Hiroko. (2013). 'Nintendo Lowers Forecast for Wii U Sales'. *New York Times.com*, 30 January, http://www.nytimes.com/2013/01/31/technology/nintendo-warns-of-weak-wii-u-sales.html?_r=0 (downloaded 23 March 2013).

Takahashi, Dean. (2004). 'Details of IBM's New Cell Chip Become Clear'. *Knight Ridder Tribune Business News*, p. 1.

Takahashi, Dean. (2005). 'Gearing up for Battle of Video Game Console Business'. *Knight Ridder Tribune Business News*, p. 1.

Takahashi, Dean. (2011). 'Global Ad Spending in Video Games to Top $7.2B in 2015'. *Venturebeat.com*, 12 September, http://venturebeat.com/2011/09/12/global-ad-spending-in-video-games-to-top-7-2b-in-2016/ (downloaded 23 March 2012).

Take-Two. (2008a). *Annual Report*. New York: Take-Two Interactive, Inc., http://ir.take2games.com/bios.cfm?pg=58 (downloaded 22 July 2008).

Take-Two. (2008b). 'Board of Directors'. New York: Take-Two Interactive, Inc., http://ir.take2games.com/bios.cfm?pg=58 (downloaded 22 July 2008).

Take-Two. (2008c). 'Statement to Shareholders: Take-Two Interactive Software, Inc. Responds to Electronic Arts' Extension of Tender Offer', http://www.taketwovalue.com/home.php (downloaded 22 July 2008).

Takiff, Jonathan. (1992). 'Test Your Olympic Videothalon Mettle'. *Philadelphia Daily News*, 29 July, p. 31.

Tallon, Mary. (2005). 'Ill. Moves toward Banning Some Video Games'. *Associated Press*, 10 March, http://news.yahoo.com/news?tmpl=story&u=/ap/20050310/ap_on_hi_te/video_game_ban (accessed 10 March 2005).

Tapscott, Don. (1997). *Growing up Digital: The Rise of the Net Generation*. New York: McGraw-Hill.

Tassi, Paul. (2011). 'Analyst Says Video Game Advertising Will Double by 2016'. *Forbes.com*, http://www.forbes.com/sites/insertcoin/2011/09/14/analyst-says-video-game-advertising-will-double-by-2016/ (downloaded 13 June 2012).

Taub, Eric A. (2005). 'HDTV Is a New Reality for Game Developers'. *New York Times*, 16 May, p. C.12.

Tedesco, Theresa. (2009). 'iPhone Economy Taking Shape: Apps World'. *National Post*, p. FP.1. http://proquest.umi.com/pqdweb?did=1747134321&Fmt=7&clientId=5258&RQT=309&VName=PQD (accessed 13 August 2013).

Thegameconsole.com. (2010). 'A Brief History of the Video Game Console', http://www.thegameconsole.com/ (downloaded 11 May 2010).

Thier, Dave. (2012). Facebook Has a Billion Users and a Revenue Question'. *Forbes.com*, 4 October, http://www.forbes.com/sites/davidthier/2012/10/04/facebook-has-a-billion-users-and-a-revenue-question/ (downloaded 23 March 2013).

Thompson, Lynn. (1994). 'Atari Ships Tempest 2000 for Jaguar'. *Business Wire*, p. 1.

Thorsen, Tor. (2003). 'Game Retailers Announce New ID-checking Plan'. *GameSpot*, 8 December, http://www.gamespot.com/articles/game-retailers-announce-new-id-checking-plan/1100-6085233/ (accessed 13 August 2013).

Thorsen, Tor. (2009). 'Midway Gets $33 Million Warner Bros. Bid'. *Gamespot.com*, 8 December, http://www.gamespot.com/news/midway-gets-33-million-warner-bros-bid-6210091 (downloaded 23 March 2013).

THQ. (2004). *Annual Report*. Agoura Hills, CA: THQ, Inc.

Thurber Jr, Karl T. (1995). 'Buried Bytes: A History of the Personal Computer'. *Popular Electronics* vol. 12 no. 4, April, p. 36.

Time. (1990). 'Dr. Nintendo'. *Time*, 28 May, p. 72.

Time Warner. (2011). *Time Warner Inc's Worldwide Subsidiaries and Affiliated Companies List*. New York: Time Warner, Inc.

Tito, Gregg. (2010). 'Study Claims Average Game Budget Is $23 Million'. *Escapist*, 12 January, http://www.escapistmagazine.com/news/view/97413-Study-Claims-Average-Game-Budget-Is-23-Million (downloaded 13 March 2010).

Toles, Terri. (1985). 'Video Games and American Military Ideology'. In Vincent Mosco and Janet Wasko (eds). *The Critical Communications Review*, Vol. III: Popular Culture and Media Events. Norwood, NJ: Ablex Publishing, pp. 207–23.

Totilo, Stephen. (2010). 'Rockstar Responds to "Rockstar Spouse" Controversy, "Saddened" by Accusations'. *Kotaku.com*, 21 January, http://kotaku.com/5452809/rockstar-responds-to-rockstar-spouse-controversy-saddened-by-accusations (downloaded 10 October 2012).

Tramain, Steve. (2002). 'Game Tie-ins with Films on the Rise'. *Billboard* vol. 114 no. 24, 15 June, p. 80.

Tran, Khan T. L. (2002). 'The Game Machine That Multitasks – New Add-ons Turn Devices into E-mail Hubs, Even PCs; Karaoke with Your Game Boy'. *Wall Street Journal*, 27 August, p. D.1.

Turley, Jim. (2006). 'A Glimpse inside the Cell Processor – What Has 250 Million Transistors and Nine Processors? *Embedded Systems Design* vol. 19 no. 6, p. 39.

Turner, Lucy. (1997). 'It's the One and Sony: PlayStation Is No1 Xmas Toy'. *Mirror*, p. 2, http://proquest.umi.com/pqdweb?did=110160300&Fmt=7&clientId=5258&RQT=309&VName=PQD (accessed 13 August 2013).

USA Today. (2004). 'Media Giants Suit up to Take on Video Games'. *USA Today*, 27 August, p. 5-B.

USK. (2007). 'The Five Ratings and What They Mean', http://translate.google.com/translate?u=http%3A%2F%2Fwww.usk.de%2F&hl=en&ie=UTF8&sl=de&tl=en (downloaded 8 July 2008).

USK. (2010). 'USK-Unterhaltungssoftware Selbstkontrolle' http://www.usk.de/ (downloaded May 2010).

Valve. (2011). 'Valve: Company', http://www.valvesoftware.com/company/ (downloaded 25 February 2011).

Van Slyke, Brandon. (2008). 'How a Video Game Gets Made'. *Game Developer 2008 Game Career Guide*, pp. 19–24.

Vara, Vauhini and Stu Woo. (2012). 'San Francisco's Pursuit of Tech Spurs Questions: Emphasis on Internet Sector Raises Concerns about Less Skilled Workers, Cost of Living. *Wall Street Journal (Online)*, 1 May, http://online.wsj.com/article/SB1000 14240527023043312045773561838625 13016.html (downloaded 13 April 2013).

Vargas, Jose Antonio. (2006). 'Brothers of Reinvention: Long after What Seemed to Be Luigi and Mario's Heyday, the Nintendo Classic Keeps Proving It's Still Got Game'. *Washington Post*, p. N.12, http://proquest.umi.com/pqdweb?did= 1089821121&Fmt=7&clientId=5258&RQT=309&VName=PQD (accessed 13 August 2013).

Veiga, Alex. (2004). ' "BitTorrent" Gives Hollywood a Headache'. *Associated Press*, 10 December, http://usatoday30.usatoday.com/tech/news/techpolicy/2004-12-10-bittorrent-hollywood_x.htm (accessed 13 August 2013).

VGChartz. (2008a). 'American Chart for 2007', http://vgchartz.com/hwtable. php?cons%5B%5D=Wii&cons%5B%5D=PS3&cons%5B%5D=X360&cons %5B%5D=XB&cons%5B%5D=GC&cons%5B%5D=PS2&cons%5B%5D= PS&reg%5B%5D=Total&start=32621&end=39628 (downloaded 8 July 2008).

VGChartz. (2008b). 'Hardware Table, Total Units 4/22/89-12/29/08', http://vgchartz. com/hwtable.php?cons%5B%5D=Wii&cons%5B%5D=PS3&cons%5B%5D= X360&cons%5B%5D=XB&cons%5B%5D=GC&cons%5B%5D=PS2&cons %5B%5D=PS&reg%5B%5D=Total&start=32621&end=39628 (downloaded 8 July 2008).

VGChartz. (2010a). 'Hardware Annual Summary', http://www.vgchartz.com/ hw_annual_summary.php (downloaded 19 September 2010).

VGChartz. (2010b). 'Hardware Comparison Table, 16 April 1989 to 25 September 2010', http://www.vgchartz.com/hwtable.php?cons%5B%5D=PSP&cons %5B%5D=DS&cons%5B%5D=GBA&cons%5B%5D=GB&reg%5B%5D=Total &reg%5B%5D=America&reg%5B%5D=Japan&reg%5B%5D=Total+Others& reg%5B%5D=UK&reg%5B%5D=France&reg%5B%5D=Germany&reg%5B %5D=Spain&reg%5B%5D=Italy&reg%5B%5D=Scandinavia&reg%5B%5D= Australia&start=32621&end=40447 (downloaded 29 Septermber 2010).

VGChartz. (2010c). 'Hardware Totals', http://www.vgchartz.com/hardware_totals.php (downloaded 19 September 2010).

VGChartz. (2010d). 'VGChartz – Hardware Comparison Table', http://www.vgchartz. com/ (downloaded 13 July 2010).

VGChartz. (2011a). 'Hardware Annual Summary', http://www.vgchartz.com/ hw_annual_summary.php (downloaded 6 February 2011).

VGChartz. (2011b). 'Software Totals: Total Worldwide Sales (in Millions of Units) per Game', http://www.vgchartz.com/ (downloaded 11 December 2011).

VGChartz. (2013a). 'Global Yearly Chart 2010', http://www.vgchartz.com/yearly/2013/Global/ (downloaded 23 March 2013).

VGChartz. (2013b). 'Global Yearly Chart, 2011', http://www.vgchartz.com/yearly/2013/Global/ (downloaded 23 March 2013).

VGChartz. (2013c). 'Global Yearly Chart 2012', http://www.vgchartz.com/yearly/2013/Global/ (downloaded 23 March 2013).

VGChartz. (2013d). 'Global Yearly Chart 2013', http://www.vgchartz.com/yearly/2013/Global/ (downloaded 23 March 2013).

Vick, Karl. (2001). 'Vital Ore Funds Congo's War; Combatants Profit from Col-Tan Trade'. *Washington Post*, 19 March, p. 0–A.1.

Vogel, Harold L. (2004). *Entertainment Industry Economics: A Guide for Financial Analysis*, 6th edn. Cambridge: Cambridge University Press.

Wadhams, Nick. (2005). 'Troops Stationed in Iraq Turn to Gaming'. *Associated Press*, 2 January, http://apnews.excite.com/article/20050103/D87CAVO80.html (downloaded 17 January 2005).

Wahl, Andrew. (2005). 'On Technology – My Pick: EA, by about a Billion'. *Canadian Business* vol. 78 no. 3, pp. 17–18.

Wakabayashi, Daisuke. (2009). 'Sony Wagers Online Push Will Spur Sales.' *Wall Street Journal Asia*, p. 1, http://proquest.umi.com/pqdweb?did=2076092831&Fmt=7&clientId=5258&RQT=309&VName=PQD (accessed 13 August 2013).

Wakabayashi, Daisuke and Yukari Iwatani Kane. (2010). 'New Sony Gadgets Take Aim at Apple'. *Wall Street Journal*, 5 March, p. B.1, http://online.wsj.com/article/SB10001424052748703502804575101013088128250.html (downloaded 2 April 2010).

Waldman, Dan. (2006). 'Pipeline Progess Family-friendly Games Fuel Digital Distribution Growth, *KidScreen*, p. 31.

Wall Street Journal. (2002). 'Business Brief – Sega Corp.: Alliance Is Formed with ESPN to Improve Sports Video Games'. *Wall Street Journal*, 9 May, p. 1.

Wallace, Mark. (2005). 'The Game Is Virtual, The Profit Is Real'. *New York Times*, 25 May, p. 3.7.

Wallace, Richard. (2004). 'Game Design Gets Serious for Real World Apps'. *Electronic Engineering Times* no. 1314, http://www.eetimes.com/document.asp?doc_id=1149764 (accessed 13 August 2013).

Wapshott, Tim. (1999). 'A Potted History of the Console.' *The Times*, p. 10, http://proquest.umi.com/pqdweb?did=45524943&Fmt=7&clientId=5258&RQT=309&VName=PQD (accessed 13 August 2013).

Wasko, Janet. (1994). *Hollywood in the Information Age*. Austin: University of Texas Press.

Wasko, Janet. (1998). 'Challenges to Hollywood's Labor Force in the 1990s'. In
    Gerald Sussman and John A. Lent (eds). *Global Productions: Labor in the Making of
    the 'Information Society'*. Creskill, NJ: Hampton Press, Inc, pp. 173–89.

Wasko, Janet. (2003). *How Hollywood Works*. London: Sage.

Watson, Rob. (2008). 'Sony's PlayStation 3 Clearly Found Its Place'. *McClatchy –
    Tribune News Service*.

Wattenburger, Phil. (2005). Telephone Interview. Austin, Texas.

Waugh, Eric-Jon. (2005). 'E3 Report: Smart Marketing – How an Intelligent Approach
    to Research Can Boost Your Bottom Line'. *Gamasutra.com*, 24 May, http://www.
    gamasutra.com/features/20050524/waugh_pfv.html (accessed 13 August 2013).

Wauters, Robin. (2011). 'Zeebo Raises $17 Million for Interactive, Entertainment,
    Education Platform'. *Techcrunch.com*, 29 August, http://techcrunch.com/
    2011/08/29/zeebo-raises-17-million-for-interactive-entertainment-education-
    platform/ (downloaded 10 October 2012).

Wayne, Teddy. (2010). 'Women Set the Pace as Online Gamers'. *New York Times*,
    8 August, http://www.nytimes.com/2010/08/09/technology/09drill.html
    (downloaded 17 August 2010).

Webster, Nancy Coltun and Beth Snyder Bulik. (2004). 'Now down to Business:
    Counting Gamer Thumbs'. *Advertising Age* vol. 75, 24 May, pp. S6–S7.

Wheelwright, Holly. (1990). 'Video Games That Grown Ups Will Love'. *Money*
    vol. 19, pp. 32–3, 38–9.

Wiegner, Kathleen K.. (1979). 'The Micro War Heats up'. *Forbes* vol. 124, pp. 49–52,
    54, 56, 58.

Williams, Dmitri. (2002). 'Structure and Competition in the U.S. Home Video Game
    Industry'. *JMM – International Journal on Media Management* vol. 4 no. 1, pp. 41–54.

Williams, Dmitri. (2003a). 'The Video Game Lightning Rod: Constructions of a New
    Media Technology, 1970–2000'. *Information, Communication, and Society* vol. 6
    no. 4, pp. 523–50.

Williams, Eric. (2003b). 'Environmental Impacts in the Production of Personal
    Computers'. In Ruediger Kuehr and Eric Williams (eds). *Computers and the
    Environment: Understanding and Managing Their Impacts*. Boston, MA: Kluwer
    Academic Press, pp. 41–72.

Wilson, Mark. (2011). 'Will Xbox 360 Replace Your Cable Box – Or Will a TV
    Replace Your Xbox?' *Popular Mechanics*, http://www.popularmechanics.com/
    technology/gadgets/home-theater/will-xbox-360-replace-your-cable-box-or-will-a-tv-
    replace-your-xbox-6612417 (accessed 13 August 2013).

Wilson, Trevor. (2006). 'Top 20 Publishers'. *Game Developer* vol. 13, pp. 11–22.

Wingfield, Nick. (2005). 'The Next Wave of Videogames: Software for New Machines
    Will Look Better, Cost More, Dismember Bodies like "Jaws"'. *Wall Street Journal*,
    12 May, p. D.1.

Wingfield, Nick and Merissa Marr. (2005). 'Games Players Can't Refuse?' *Wall Street Journal*, 19 April, p. B.1.

Wired. (2003). 'TV Execs Go Gaga over Gaming.' *Wired.com*, 17 December, http://www.wired.com/news/business/0,01367,61627,00.html (accessed 13 August 2013).

Wise, Deborah C. (1985). 'Tramiel's Atari: The Long Shot That's Coming in'. *Business Week*, 9 December, pp. 39–40.

Wolverton, Troy. (2010). 'New Platforms Open Doors to Video Game Innovation'. *Tribune – Review / Pittsburgh Tribune – Review*, http://proquest.umi.com/pqdweb?did=1982600801&Fmt=7&clientId=5258&RQT=309&VName=PQD (accessed 13 August 2013).

Wong, May. (2005). 'Gaming Industry Adopts New Tween Rating'. *Associated Press*, 3 March, http://www.highbeam.com/doc/1P1-105930321.html (accessed 13 August 2013).

Yates, Michael D. (2001). 'The "New" Economy and the Labor Movement'. *Monthly Review* vol. 52 no. 11, pp. 28–42.

Yates, Michael D. (2003). *Naming the System: Inequality and Work in the Global Economy*. New York: Monthly Review Press.

Zackariasson, Peter and Timothy L. Wilson. (2012). 'Marketing of Video Games'. In Peter Zackariasson and Timothy L. Wilson (eds). *The Video Game Industry: Formation, Present State, and Future*. New York: Routledge, pp. 57–75.

Zito, Kelly. (2000). 'Not All Fun and Games'. *Cooljobs.com*, 8 May, http://www.sfgate.com/business/article/Not-All-Fun-and-Games-Testers-get-paid-to-play-3240376.php (accessed 13 August 2013).

Index